SCIENCE
A CLOSER LOOK
Reading Essentials

Macmillan/McGraw-Hill

Macmillan/McGraw-Hill

Contents

CHAPTER 1

Plants

The Big Idea

How do plants grow and change?

Vocabulary

minerals bits of rock and soil that help living things grow

oxygen a gas that plants give off when they make food

flower plant part that makes seeds or fruit

seed plant part that can grow into a new plant

pollen a sticky powder found inside flowers

life cycle steps that show how a living thing grows, changes, and makes new living things

trait the way a plant or animal looks or acts

What do living things need?

Living things grow and change over time. Animals are living things. Animals need food, water, air, and space to live.

A swan takes care of her young. ▼

Plants are living things too. Plants also grow and change over time.

Plants need water, air, food, and space to grow. Plants make their own food.

This plant grew over time. ▶

sprout

young plant

adult plant

![check] **Quick Check**

I. What do living things need to live?

_____

_____

How do plants make food?

Plants have special parts they use to make food. Plants need sunlight, air, water, and minerals to make food. **Minerals** are bits of rock and soil. Minerals help plants and animals grow.

Plants Make Food

Leaves use air and sunlight to make food.

Water and food move through the stem.

Roots hold plants in the soil. They take in water and minerals.

Read a Diagram

What holds the plant in the soil?

_____

Plants give off a gas called **oxygen** when they make food. All animals need oxygen to live.

Humans and other animals breathe oxygen. ▼

 Quick Check

Fill in the blanks.

2. Plants use minerals, air, water, and _____ to make food.

3. Plants need _____ from the soil to grow.

Where do seeds come from?

A flower is a part of a plant. A **flower** is the part that makes seeds, fruit, and pollen. Seeds are parts of plants too. **Seeds** grow into new plants. Fruit protects the seeds.

Part of a flower makes pollen. **Pollen** is sticky powder inside a flower. It helps make seeds.

Making a Melon

1. This part of the flower makes pollen.

2. This flower part grows into a fruit with seeds.

Animals can move pollen from flower to flower. Wind and water can move pollen too.

 Quick Check

Circle the answer.

4. The _____ protects the seeds inside.

fruit   pollen

5. Wind can move _____ from flower to flower.

pollen   animals

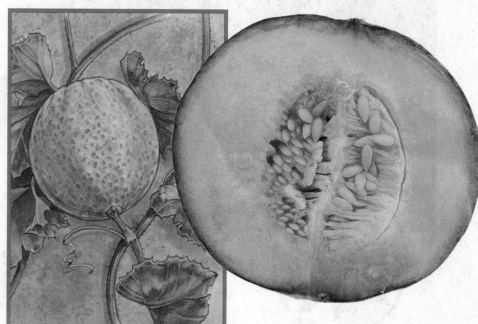

3. The fruit protects the seeds inside.

4. When the fruit is ripe, it is ready to eat.

5. The seeds inside can grow into new plants.

How do seeds look?

Seeds have different sizes and shapes. All seeds have food to help the new plant grow.

Seeds move in different ways. Wind can carry small seeds away. Other seeds stick to the fur of animals. The animals move the seeds to a new place.

star anise

marigold

star anise pod

◀ These small seeds are in a pod shaped like a star.

marigold seeds

▲ Marigold seeds are small and thin.

All seeds have covers called seed coats. Seed coats protect the seeds. Seed coats help keep the seeds from drying out.

▲ Peanuts are seeds. They grow underground.

Peanuts are protected by hard shells and seed coats. ▼

shell

seed coat

✔ Quick Check

6. Why do you think peanuts have hard shells?

_____

_____

How do seeds grow?

The **life cycle** of a living thing shows how it grows, lives, makes more of its own kind, and dies. A plant's life cycle starts with a seed. A seed needs light, water, food, and a warm place to grow.

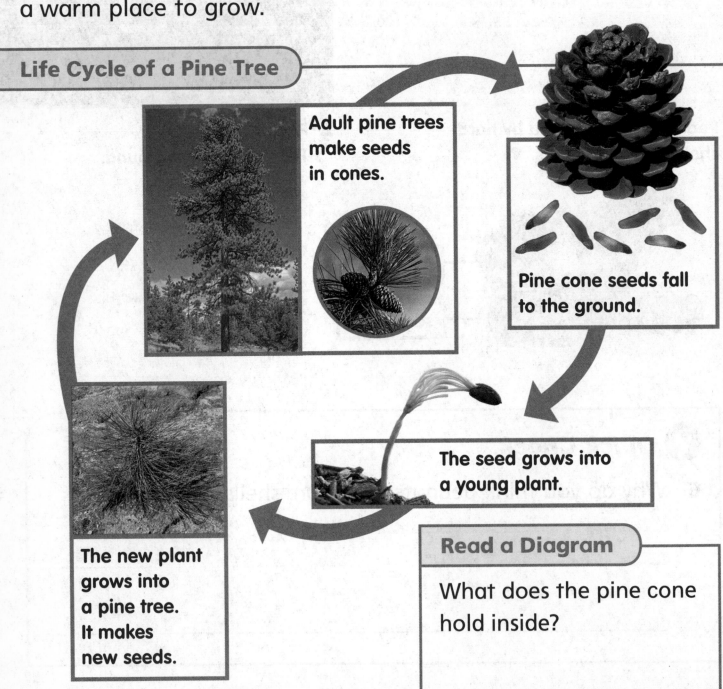

Life Cycle of a Pine Tree

Adult pine trees make seeds in cones.

Pine cone seeds fall to the ground.

The seed grows into a young plant.

The new plant grows into a pine tree. It makes new seeds.

Read a Diagram

What does the pine cone hold inside?

_____

LOG ON *Science in Motion* Watch a plant grow at www.macmillanmh.com

Different types of plants have different life cycles. A plant and its parents have the same life cycle.

▲ A redwood tree has a long life cycle.

▲ Each of these flowers has a short life cycle.

✔ Quick Check

Write *true* if the sentence is true. Write *false* if the sentence is false.

7. The life cycle of many plants begins with a seed.

_____

8. A plant has the same life cycle as its parent plants.

_____

How are plants like their parents?

Animals have babies that look and act like their parents. They have bodies that are shaped the same.

Most young plants look like their parent plants. The parts of the young plant look like the parts of the parents.

sunflower

acorn

sunflower seed

oak tree

A **trait** is the way a plant or animal looks or acts. Most young plants will have some of the same traits as their parents. Some may have different traits.

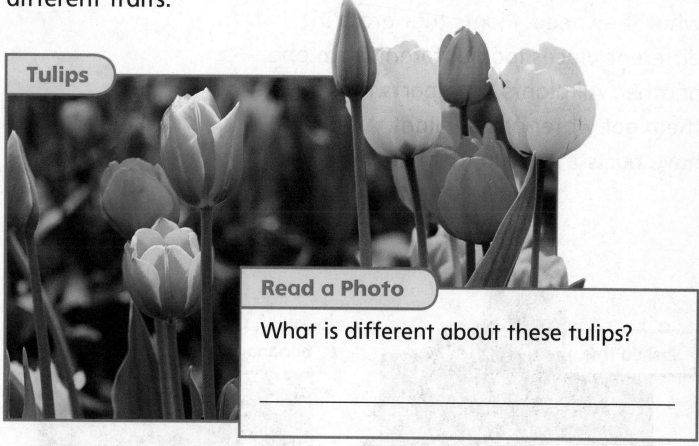

Tulips

Read a Photo

What is different about these tulips?

_____

 **Quick Check**

9. What do sunflower seeds grow up to be?

_____

10. How are young plants like their parents?

_____

How do plants survive in different places?

Plants have different ways to get what they need. Plants that grow in different places look different from one another. All plants have parts that help them get water and sunlight. All plants have parts that help them make food.

Joshua tree

▲ This tree grows in a dry place. It stores water in thick stems.

banana tree

▲ This tree grows in a wet place. The large leaves help it get light in a dark forest.

Plants can change to stay safe too. Some plants have to stay safe from animals. Other plants have traits that help them stay safe from the weather.

▲ **This tree grows sideways because of strong winds.**

 Quick Check

Fill in the blanks.

11. Plants may have different shapes,

but their parts _____.

12. Plants may have to stay

safe from animals and the _____.

Use the words in the box to fill in the blanks.

flower	minerals	pollen
life cycle	oxygen	seed

1. The sticky powder inside a flower is called

_____ .

2. Plants give off _____ that animals breathe.

3. A _____ can grow into a new plant.

4. The _____ is the part of the plant that makes seeds.

5. A _____ shows how a living thing grows.

6. Bits of rock and soil that help living things grow are called

_____ .

Write what you learned.

_____

_____

CHAPTER 2

Animals

The Big Idea

How do animals grow and change?

Vocabulary

mammal an animal with hair or fur that feeds milk to its babies	
reptile an animal with dry, scaly skin	
amphibian an animal that lives part of its life in water and part on land	
insect an animal with six legs and an outer shell	
larva a stage in the life cycle of some animals after they hatch from an egg	
pupa the third stage in an insect's life cycle before it changes into an adult	

How do we group animals?

Scientists can sort animals into two main groups. The animals in one group have backbones. Animals in the other group do not have backbones.

Animals with backbones can be sorted into smaller groups. **Mammals** are animals that have hair or fur. **Reptiles** have dry skin with scales.

alligator

▲ **Most reptiles hatch from eggs.**

Mammals feed milk to their young. ▼

lions

Birds, fish, and amphibians are other animal groups with backbones.

Birds have feathers and lay eggs. Fish live in water. They breathe with gills and swim with fins. **Amphibians** live in water and on land. They have moist skin.

bluebird

▲ Most birds can fly.

▼ **Many fish lay eggs and have scales.**

salmon

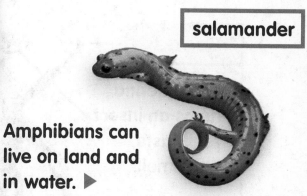

salamander

Amphibians can live on land and in water. ▶

✔️ *Quick Check*

Fill in the blank.

1. Animals that swim with fins are called _____.

2. Animals that have fur or hair are called _____.

What are some animals without backbones?

Animals without backbones can also be sorted into smaller groups. Some have shells or hard body coverings that help them stay safe.

Insects do not have backbones. **Insects** have six legs and an outer shell.

blue crayfish

◀ Claws help keep this animal safe.

Beetle

The antenna helps an insect feel, taste, and smell.

A hard outer shell helps keep an insect safe.

Read a Diagram

What part helps keep the beetle safe?

_____

Other animals without backbones do not have shells. Their bodies are soft. These animals have other ways to stay safe.

Jellyfish sting other animals to stay safe. ▶

jellyfish

 Quick Check

What are two ways animals without backbones can stay safe?

3. _____

4. _____

What is a life cycle?

All animals have a life cycle. Different kinds of animals have different life cycles. Birds and reptiles hatch from eggs.

Chicken Life Cycle

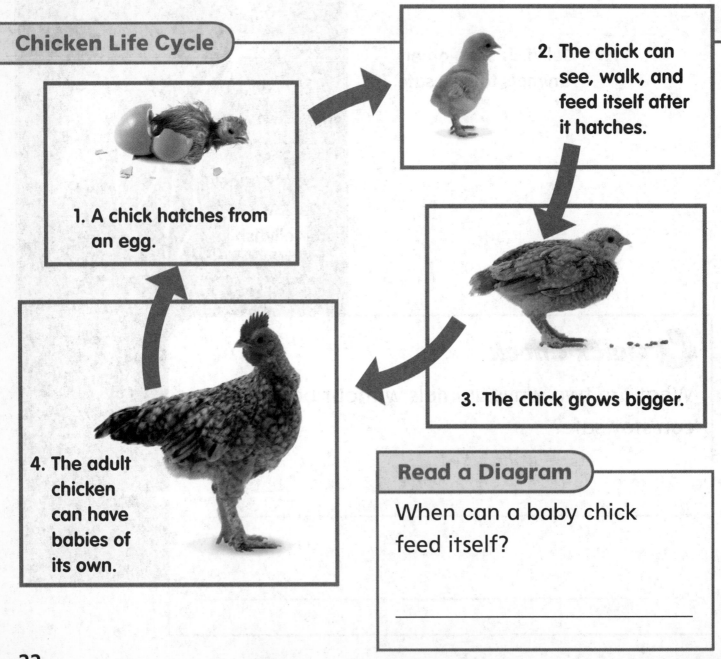

1. A chick hatches from an egg.

2. The chick can see, walk, and feed itself after it hatches.

3. The chick grows bigger.

4. The adult chicken can have babies of its own.

Read a Diagram

When can a baby chick feed itself?

_____

LOG ON *Science in Motion* Watch animals grow at www.macmillanmh.com

Mammals give birth to live babies. These babies grow into adults. Adults can have babies of their own.

adult panda with young panda

baby panda

▲ Pandas are a kind of mammal. They give birth to live babies.

 Quick Check

Write *true* if the sentence is true. Write *false* if the sentence is false.

5. All animals have the same life cycle.

_____

6. Most mammals hatch from eggs.

_____

What are some other animal life cycles?

Some animals do not look like their parents when they are born. They change as they grow.

Animals like butterflies and frogs grow in stages. Butterflies begin as eggs. The egg hatches into a **larva**. This is the second stage. The larva changes into a **pupa** in the third stage. Then the pupa changes into an adult. This is the last stage.

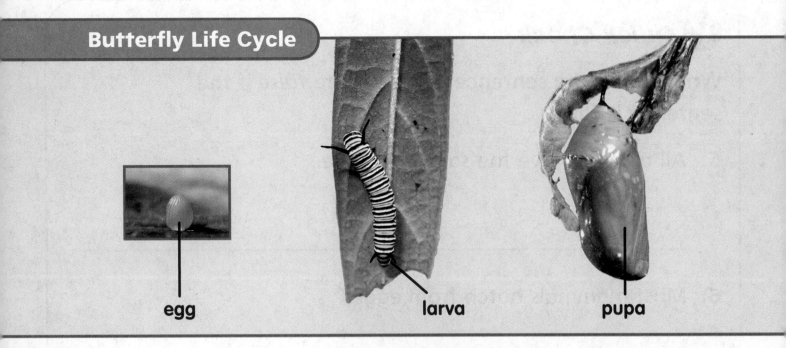

Butterfly Life Cycle

egg

larva

pupa

▲ The larva of a butterfly is called a caterpillar. The caterpillar looks very different from an adult butterfly.

Quick Check

Circle the answer.

7. A butterfly egg hatches into a(n) _____.

adult larva pupa

8. An adult butterfly comes from the _____.

egg larva pupa

hatching adult   adult

Why do animals act and look the way they do?

Animals act and look the way they do to help them stay alive. Some animals have body parts that help them find food. Some animals have colors, patterns, or shapes that help them hide from other animals.

giraffe

◄ A long neck helps a giraffe eat leaves in tree tops.

tarsier

Big eyes help this animal see at night. Long fingers help it dig for food. ▶

snow leopard

A leopard's spots make it hard to see against the rocks. ▶

Bird Feathers

In summer this bird has brown feathers.	In fall the bird's feathers begin to turn white.	In winter its feathers blend with the snow.

Read a Photo

What color is this bird in winter?

_____

 Quick Check

9. How do an animal's body parts help it stay safe or get food?

_____

_____

_____

How do animals stay safe?

Animals have different ways to stay safe. Some animals stay in large groups. Some leave their homes in winter to be in a warm place with food.

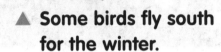

sandhill cranes

▲ Some birds fly south for the winter.

dormouse

◄ This animal sleeps during the cold winter.

▼ These fish swim in a large group to stay safe.

yellow goatfish

Animals may have body parts to keep them safe. Some have shells or smells to protect them from other animals.

turtle

▲ Turtles can hide in their shells.

skunk

▲ Skunks spray a bad smell to keep other animals away.

✔ Quick Check

Write how each animal can stay safe.

10. A turtle can stay safe when it _____.

11. Fish can stay safe when they _____.

Draw a line from each word to its meaning.

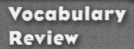 pupa

1. an animal with hair or fur that makes milk for its babies

 insect

2. an animal with six legs and an outer shell

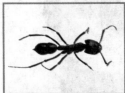 mammal

3. an animal with dry, scaly skin

 larva

4. the stage in the life cycle of some animals after they hatch from an egg

 reptile

5. the third stage in an insect's life cycle before it changes into an adult

Write what you learned.

_____

_____

CHAPTER 3

Looking at Habitats

The Big Idea

What are habitats?

Vocabulary

habitat a place where plants and animals live

food chain a model that shows the order in which living things get the food they need

food web two or more food chains that are connected

endangered when few of one kind of animal are left

fossil what is left of a living thing from the past

extinct when a living thing dies out and no more of its kind live on Earth

What is a habitat?

A **habitat** is a place where plants and animals live. Plants and animals find what they need to live in their habitats.

buffalo

◀ **grassy and warm**

red-necked grebe

wet and grassy ▶

There are many kinds of habitats. Some are rainy and some are dry. Some are hot and some are cold. Different plants and animals need different habitats to live.

tortoise

▲ hot and dry

polar bears

▲ cold and snowy

✓ Quick Check

I. List some different types of habitats.

_____

_____

_____

How do living things use their habitats?

Animals use plants in their habitats for food. Some animals eat other animals that live in their habitats. Animals can hide and sleep in their habitats.

Forest Habitat

Read a Diagram

What part of this habitat does the squirrel use for shelter?

_____

Different plants need different habitats. Some plants need sandy soil. Some plants grow in rocky soil. Some plants live in dry places. Some plants live where it is very wet.

This plant lives in a dry place. Its leaves store water. ▶

Quick Check

Name three ways animals use their habitats.

2. _____

3. _____

4. _____

What is a food chain?

A **food chain** shows the order in which living things get the food they need. Food chains can be on land, in water, or both.

Most food chains start with the Sun. Plants use sunlight to make food. Some animals eat plants. Some animals eat other animals.

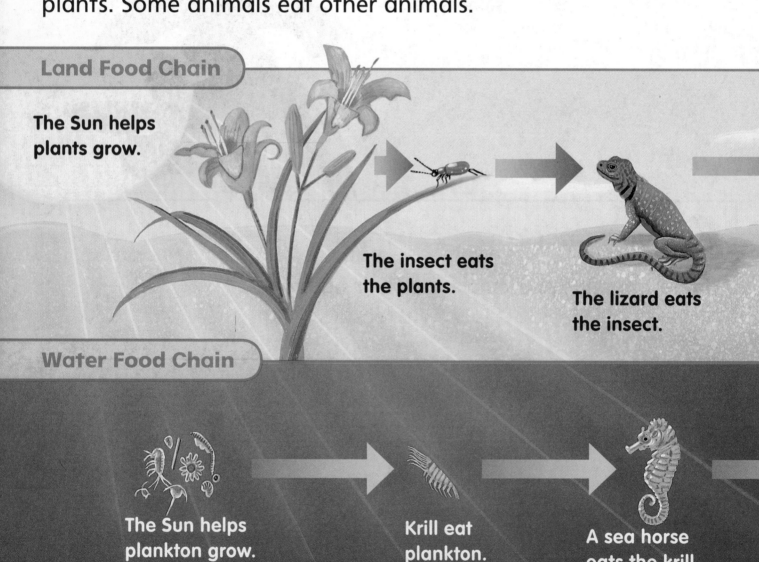

Land Food Chain

The Sun helps plants grow.

The insect eats the plants.

The lizard eats the insect.

Water Food Chain

The Sun helps plankton grow.

Krill eat plankton.

A sea horse eats the krill.

Some animals eat things that are dead. They break up the dead things into small pieces.

 Quick Check

Circle the answer.

5. The first living thing in a land food chain is _____.

a plant   the Sun   an insect

6. Some animals eat things that are _____.

pieces   dead   chains

The snake eats the lizard.

The hawk eats the snake.

Large fish eat sea horses.

Sharks eat large fish.

A Desert Food Web

Arrows in the food web go from food to eater.

Read a Diagram

What animal is eaten by the lizard and the frog?

_____

LOG ON *Science in Motion* See the parts of a food web at www.macmillanmh.com

What is a food web?

A **food web** is two or more food chains that are connected. One kind of plant or animal can be food for many animals. Mice are food for hawks, owls, and snakes.

Animals also eat more than one kind of animal. Hawks eat mice, rabbits, frogs, and snakes.

This bird is eating an insect. Together their food chains make a food web. ▶

✅ **Quick Check**

Fill in the blanks.

7. If you put together two or more food chains, you

   have a _____.

8. Hawks, owls, and snakes all eat _____.

How do habitats change?

Habitats can change in many ways. A drought is a period of time with little or no rain. A drought can change a habitat. Floods and fires can also change habitats.

Animals can change habitats. Beavers make dams. The dam can make a pond.

Drought

Read a Photo

What do you see that needs water?

_____

beaver dam

People also change habitats. People build houses, buildings, and roads where plants and animals live.

▲ Building houses and roads changes a habitat.

▲ Clearing the land changes a habitat.

 Quick Check

Write *true* if the sentence is true. Write *false* if the sentence is false.

9. Only humans can change a habitat. _____

10. Floods and fires can change a habitat. _____

What happens when habitats change?

When habitats change, animals may not be able to find the things they need. Sometimes animals can change or find a new place to live.

tigers

▲ People hunt tigers and cut down their forest homes.

whooping crane

People build homes where whooping cranes live. ▶

Some animals may die when their habitats change. When few of one kind of animal are alive, that animal is **endangered**. Animals can also become endangered when people hunt them.

manatee

▲ Fishing nets and powerboats hurt manatees.

 Quick Check

11. What can happen to an animal when its habitat changes?

_____

12. How can people make an animal endangered?

_____

How can we tell what a habitat used to be like?

A **fossil** is what is left of a living thing from the past. Scientists use fossils to learn about the past.

Some fossils do not match the habitat where they are found. This tells scientists that the habitat has changed.

When scientists found this fossil, they knew this habitat was different long ago. ▶

fossil

Some plants and animals that lived long ago still live today. Plants and animals that have died out are **extinct**. Their fossils tell us how they may have looked.

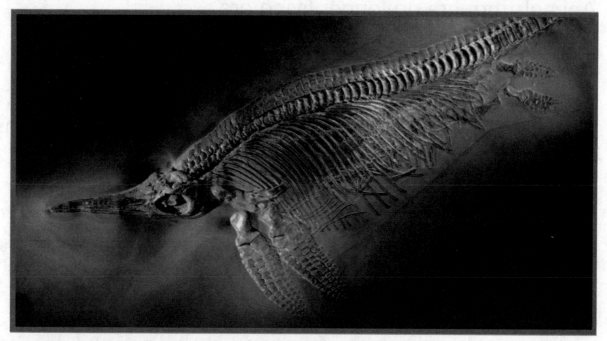

▲ This fossil is from an animal that is now extinct. It shows the shape of the animal.

✔ *Quick Check*

Fill in the blanks.

13. Animals that have died out are _____.

14. A living thing from the past may leave behind a(n)

_____.

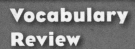

Circle the answer.

1. Animals that are no longer living are _____.

   mammals   extinct   fish

2. A _____ is two or more connected food chains.

   food web   habitat   fossils

3. A _____ is what is left of a living thing from the past.

   extinct   food web   fossil

4. A place where plants and animals get what they need to live
   is called a _____.

   fossil   habitat   food web

Write what you learned.

_____

_____

CHAPTER 4

Kinds of Habitats

What are different kinds of habitats?

Vocabulary

woodland forest a habitat where trees can grow well	
rain forest a habitat where it rains almost every day	
desert a dry habitat that gets very little rain	
Arctic a very cold place near the North Pole	
ocean a large body of salty water	
pond a small body of fresh water	

What is a woodland forest like?

A **woodland forest** is a habitat that gets plenty of rain and sunlight. Trees grow well here. Some trees have leaves that turn color and drop in the fall. Other trees stay green all year. Many different animals and plants live here.

Animals in a Woodland Forest

woodchuck

white-tailed deer

Forest animals stay alive in different ways. There is a lot to eat during the spring, the summer, and the fall. In the winter, there is not much food. Some animals sleep for the winter.

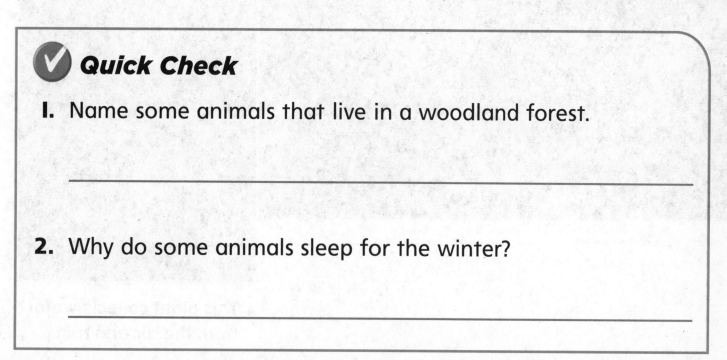

Woodland Animals

Owls have large eyes to hunt at night.

Woodpeckers have sharp beaks to find insects in trees.

Deer have colors and spots that help them hide in the forest.

Read a Chart

What body part helps owls hunt at night?

_____

✔ Quick Check

1. Name some animals that live in a woodland forest.

_____

2. Why do some animals sleep for the winter?

_____

What is a tropical rain forest?

A tropical **rain forest** is a habitat. It rains there almost every day.

Many plants live in different parts of the rain forest. Trees grow very tall. Their large leaves can block sunlight. Some plants live on the trees to get more sunlight.

rain forest

orchid

▲ **This plant collects water from the air and rain.**

Animals live in different parts of the rain forest too. Some animals live in the treetops. Other animals live on the ground.

macaw

▲ This bird eats fruit and seeds from high in the trees.

tapir

▲ This animal hunts for food on the ground.

✓ **Quick Check**

Complete each sentence.

3. An animal that gets its food from the rain forest trees is

_____.

4. The rain forest trees are

_____.

What is a hot desert like?

A **desert** is a habitat that is dry. A desert gets very little rain. A hot desert is hot during the day and cool at night.

Desert plants can store water in their stems and leaves. Some desert plants have special roots. These roots spread out wide or deep to find water when it rains. Other plants have leaves that curl up to hide from the sunlight.

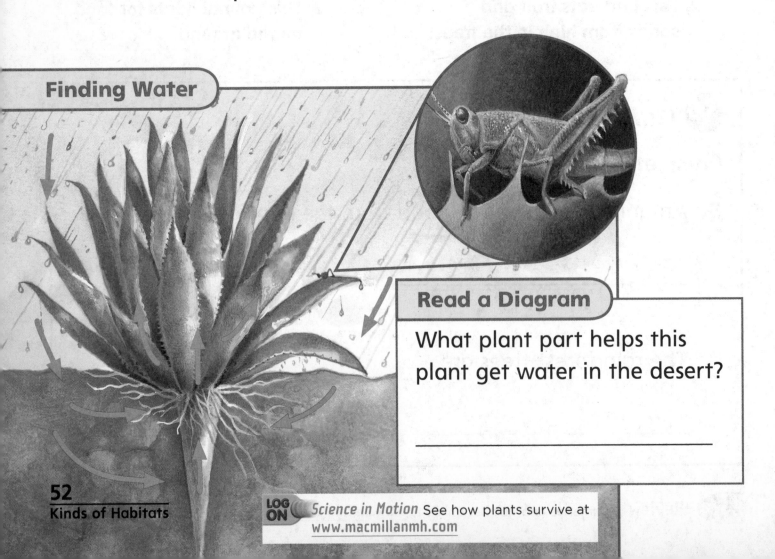

Finding Water

Read a Diagram

What plant part helps this plant get water in the desert?

_____

LOG ON *Science in Motion* See how plants survive at
www.macmillanmh.com

Desert animals live with little water. They get most of their water from eating plants or other animals.

Most desert animals sleep during the day when it is very hot. They come out at night when it is cool. Other animals have pale colors that help them stay cool in the sunlight.

snake

Gila monster

coyote

 Quick Check

Complete each sentence.

5. Some desert plants can store water in their

_____.

6. To keep cool, desert animals

_____.

What is the Arctic like?

The **Arctic** is a habitat near the North Pole. It is a cold desert.

Arctic animals stay warm in different ways. Some animals grow thick fur. Other animals have thick blubber to keep them warm.

▲ **Walruses have blubber to keep them warm.**

Reindeer have thick fur. ▶

There are no tall trees in the Arctic. Plants grow very low to the ground. This helps protect the plants from cold winds.

Arctic plants have short roots because the ground is frozen under the surface. ▼

✓ Quick Check

Write *true* if the sentence is true. Write *false* if the sentence is false.

7. Tall trees grow in the Arctic. _____

8. Some arctic animals grow thick fur to stay warm. _____

What is the ocean like?

The **ocean** is a large body of salt water. The water is always flowing. Oceans cover most of Earth.

Kelp is an ocean plant. Ocean animals find food and shelter in kelp.

Different animals live in different parts of the ocean. ▼

Life in the Ocean

Whales, dolphins, and sharks swim in the deep water, away from shore.

Crabs and sea stars live closer to the ocean's shore.

Read a Diagram

What lives in the deepest part of the ocean?

_____

Ocean animals have body shapes and parts to help them move in the water. They also have body parts to keep them safe.

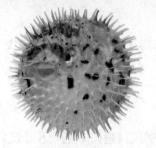

▲ Blowfish have spines to protect them.

 Quick Check

9. What is an ocean?

_____

10. Name three animals that live in the ocean.

_____

Angler fish and squid live in the deepest part of the ocean.

▼ Jellyfish sting to stay safe.

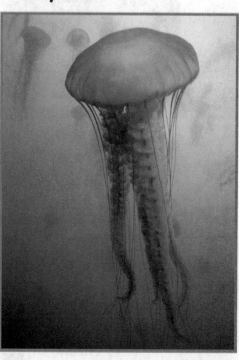

What is a pond like?

A **pond** is a small body of fresh water. The water does not move.

Pond plants grow near the shore. The stems, leaves, and flowers grow out of the water to get sunlight. Some animals eat the pond plants. Other animals eat insects, fish, frogs, and other pond animals.

little blue heron

Many plants and animals live in or near this pond. ▼

frog

muskrat

water snake

Pond animals breathe in different ways. Fish breathe with gills. Some insects breathe through a tube. There are even pond animals that breathe through their skin!

▼ Fish breathe with gills.

▲ This beetle carries a bubble of air so it can breathe under water.

✔ Quick Check

II. How can pond animals breathe?

_____

_____

Write the number of the correct meaning in front of each word.

_____ woodland forest

1. a very cold place near the North Pole

_____ ocean

2. a habitat where it rains almost every day

_____ pond

3. a habitat where it rains very little

_____ rain forest

4. a large body of salt water

_____ desert

5. a small body of fresh water

_____ Arctic

6. a habitat where lots of trees can grow well

Write what you learned.

_____

_____

CHAPTER 5

Land and Water

How can we describe Earth's land and water?

Vocabulary

crust Earth's outer layer	crust
mantle a very hot layer below Earth's crust	mantle
core Earth's deepest and hottest layer	core
earthquake a shake in Earth's crust	
volcano an opening in Earth's crust and mantle	
flood water that moves over land and does not soak into the ground	
landslide when rocks and soil slide from higher ground to lower ground	

What is land like on Earth?

Earth's land has different shapes. Land can be high or low. It can be rocky, smooth, or flat. Each of these shapes is called a **landform**.

A mountain is a high area of land. A valley is low land between mountains or hills. ▼

mountain

valley

The land is different all over Earth. Some places have lots of mountains and valleys. Other places are flat.

▲ A plain is land that is flat and wide.

▲ Hills are not as high as mountains.

 Quick Check

Fill in the blanks.

1. Earth's land has shapes that are called ___landforms___.

2. Land between mountains is called a ___Valley___.

What can maps tell us about Earth?

Maps show us where land and water are. Maps can show where the land is high or low. Look at the map below. The green parts show land that is flat. The brown parts show mountains.

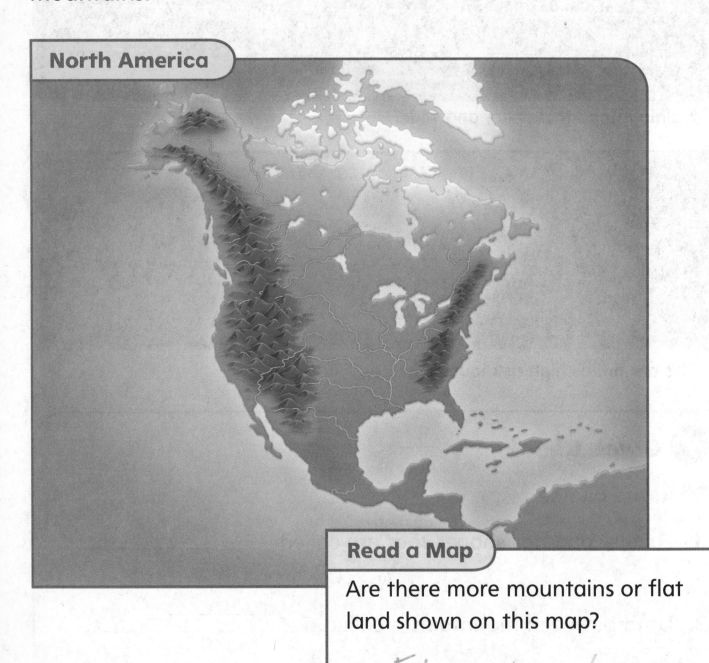

North America

Read a Map

Are there more mountains or flat land shown on this map?

Flat land

Many maps are flat, but they can also be round. A globe is a model of Earth. It is a map in the shape of a ball.

A continent is a large piece of land.

Oceans surround the continents.

✔ **Quick Check**

Circle the answer.

3. Maps can be round or _____.

mountains   (flat)   globe

4. A ball-shaped model of Earth is called a _____.

mountain   (globe)   land

What is inside Earth?

Earth is made of layers. The layer we live on is called the **crust**. It is the thinnest layer. The next layer is called the **mantle**. It is very hot. The **core** is at the center of Earth. It is the hottest part of all!

▼ We live on the top part of the crust.

crust

mantle

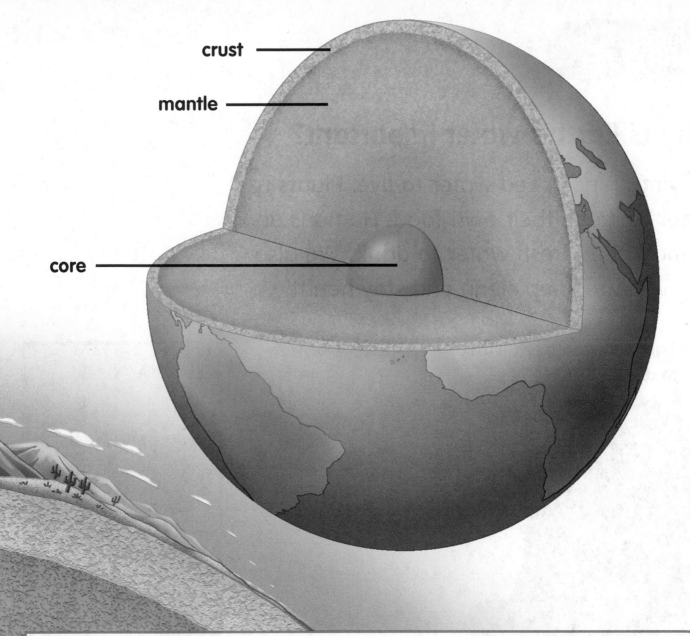

crust

mantle

core

✓ **Quick Check**

5. What are the three layers of Earth?

<u>Crust</u>   <u>mantle</u>   <u>core</u>

Why is Earth's water important?

Living things need water to live. Plants use water to make their own food. Humans and animals need fresh water to drink. We also use water to keep clean and stay healthy.

Animals need fresh water to drink. ▼

Plants get water from rain. ▶

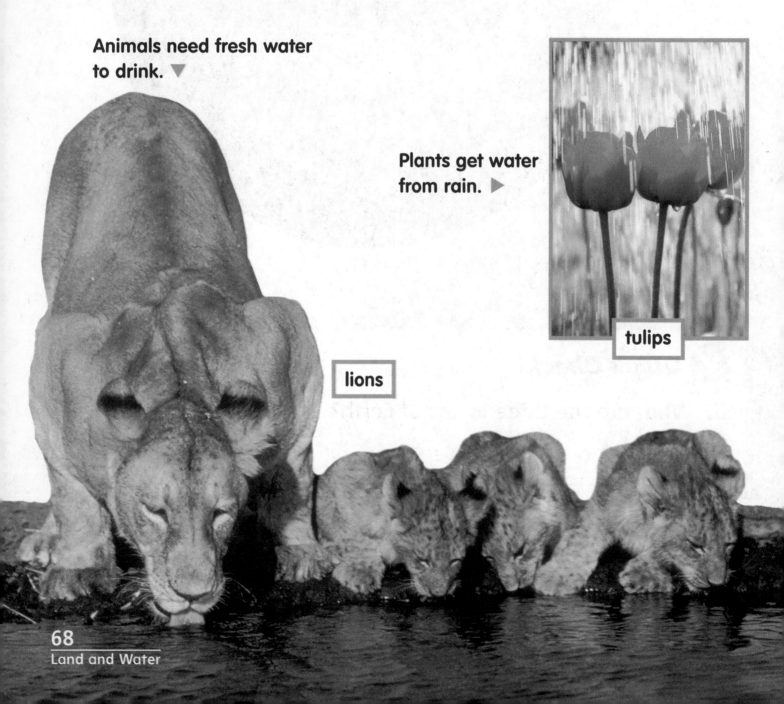

tulips

lions

Fresh water is water that is not salty. It is found in rivers, ponds, lakes, and streams.

This dam holds water. The water is cleaned. Then pipes take it into homes. ▶

✅ *Quick Check*

6. Where can we find fresh water?

lakes

streams

7. How do people use water?

swimming

drinking

Where is most of Earth's water found?

An ocean is a large, deep body of salty water. Oceans cover most of Earth. There is more water than land on Earth.

Earth from Space

Read a Photo

What do the blue parts on Earth show?

Water

Many living things live in the ocean. They need salt water to stay alive.

People use oceans. Ships carry people and goods around the world. People also get food from the ocean.

▲ **Big ships travel on the ocean.**

Dolphins live in the ocean. ▼

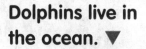

 Quick Check

Circle all correct answers.

8. How do humans use oceans?

drinking water　　　move goods　　　get food

How does Earth change slowly?

Earth changes every day. Some changes take a very long time. You can not see some of these changes.

Water can change Earth. Moving water can wear away rock. Strong wind can change Earth. Wind can blow away sand and soil.

▲ **Water shaped this canyon.**

Wind shaped these sand dunes. ▼

Ice can change Earth too. Ice can break up large rocks over a long time.

Ice Changes Land

1 Water turns to ice when it is cold. Ice takes up more space than water. The ice cracks the rock.

2 The ice melts when it is warm. There is a crack in the rock.

3 Rocks with large cracks can break apart.

Read a Diagram

What changed the rocks?

Ice

✔ Quick Check

Write *true* if the sentence is true. Write *false* if the sentence is false.

9. All Earth's changes happen quickly. _False_

10. Moving water can change the land. _Ture_

How does Earth change quickly?

Changes in Earth's crust can change the land quickly. When Earth's crust shakes, there is an **earthquake**. A **volcano** is an opening in Earth's crust. This opening can let a very hot liquid come out. The liquid cools into solid rock.

◀ Earthquakes can break apart the land.

A volcano can change Earth quickly. ▼

Other fast changes on Earth are caused by water. A **flood** can happen when rain can not soak into the ground. A **landslide** can happen when rocks and soil slide from a high place to a lower place.

Rain or melting snow can cause landslides. ▶

✔ Quick Check

II. Draw one way Earth can change quickly.

Complete each sentence with a word from the box.

crust	landform	volcano
earthquake	landslide	

1. One of the shapes of Earth's land is a(n) _landform_.

2. When Earth's crust shakes, it is a(n) _earthquake_.

3. The layer of Earth we live on is the _crust_.

4. An opening in Earth's crust is called a(n) _volcano_.

5. When rocks and soil slide from a high place to a lower place,

   it is a(n) _land slide_.

Write what you learned.

I learned that when land slides happen is when rocks and soil on high places slide

CHAPTER 6

Earth's Resources

The
Big
Idea

How do we use Earth's resources?

Vocabulary

natural resource something from Earth that people use

rock a natural resource that is hard and nonliving

minerals hard, nonliving parts of soil

soil a mix of tiny rocks and bits of plants and animals

pollution anything that makes air, land, or water dirty

litter garbage that is not thrown away

recycle to make new items out of old items

What are rocks?

A **natural resource** is something from Earth that people use. Air, water, plants, animals, and rocks are all natural resources.

A **rock** is a natural resource that is nonliving. Most rocks are hard. Rocks are everywhere on Earth. They are below grass and soil. They are even found at the bottom of the ocean!

▼ Rocks come in all shapes and sizes.

We use rocks in many ways. Rocks can be used as tools and for building. Many rocks can be carved, chipped, or ground.

▲ Rock can be carved to make statues.

▲ Long ago, people made this ax head from rock.

✓ Quick Check

1. Where can you find rocks?

_____

2. How do people use rocks?

_____

What are minerals?

Minerals are hard, nonliving parts of soil. You can find minerals in rocks. Rocks can be made of one or more minerals. People use minerals every day.

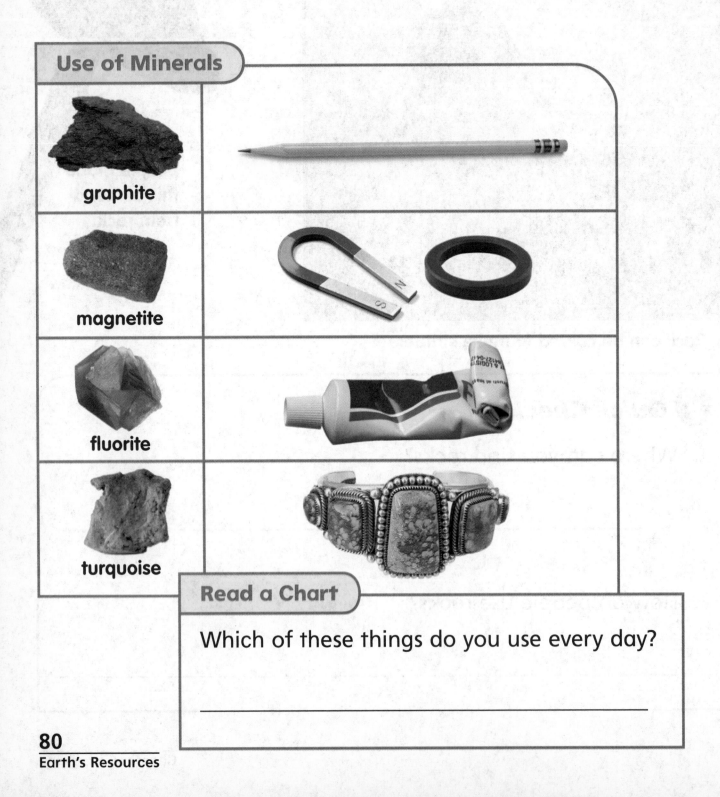

Use of Minerals

graphite	
magnetite	
fluorite	
turquoise	

Read a Chart

Which of these things do you use every day?

_____

Rocks and minerals form in the ground.
It takes millions of years for them to form.
People dig in the ground to find minerals.

A rock hammer is used to break rocks. ▶

✔ *Quick Check*

Name three ways we use minerals.

3.

Ways We
Use Minerals

What is soil?

Soil is a mix of tiny rocks and bits of plants and animals. Soil can have different colors. It can have small bits and large bits of rock.

Some soils are dark. Some hold lots of water. Soil can have large, rough pieces of rock. Other soils have rock pieces so small that they feel smooth between your fingers.

Types of Soil

clay soil

topsoil

Circle the correct answer.

4. Soil is made up of tiny rocks and bits of plants and animals.

true   false

5. All soil is the same.

true   false

sandy soil

Read a Photo

What is one way soils
can be different?

_____

How is soil formed?

Soil is formed over a very long time. Rocks and minerals break into small pieces. Dead plants and animals rot into small pieces. All of these bits become part of the soil.

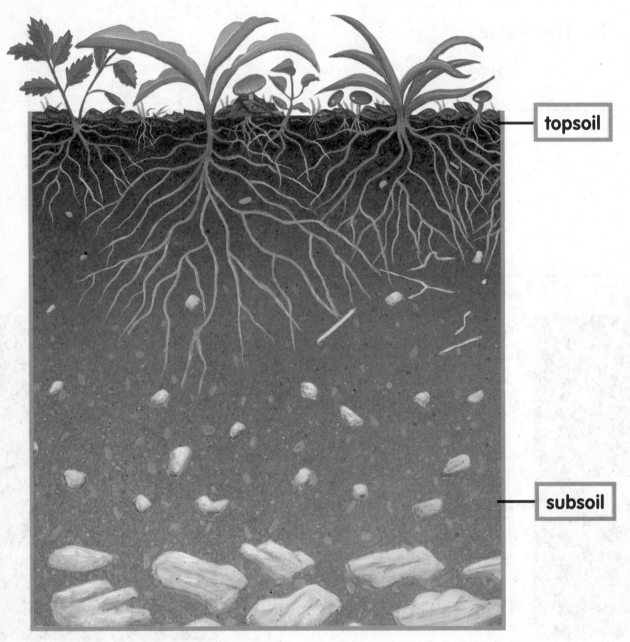

topsoil

subsoil

▲ **Plants grow best in topsoil.**

◀ This log will break down over time. It will become part of the soil.

✓ Quick Check

6. What becomes part of the soil?

_____

7. What do plant and animal parts do before they become part of the soil?

_____

How do we use natural resources?

Natural resources give us many of the things we need to live. Wind and water are natural resources we use every day. They are replaced quickly by Earth.

Moving water is used to make power. ▶

◀ **Wind can make power to light and heat homes.**

Other natural resources take a very long time to form. It can take millions of years to make some minerals! Once these are gone, they can not be quickly replaced.

Coal is a rock. It can be used to heat homes. ▶

Soil is a natural resource used to grow food. ▼

Oil comes from the ground. It is used for fuel. ▶

✔ Quick Check

8. What is one natural resource that takes a long time to replace?

_____

9. What can be used to make power?

_____

Why should we care for Earth's resources?

We need Earth's air, water, and land to live. **Pollution** is anything that makes these resources dirty.

Cars and trucks can make air pollution. Oil spills and other garbage in the water can hurt animals. **Litter** is garbage that people do not put in the right place. People can pick up litter to stop land and water pollution.

Pollution

10. How can you help stop pollution?

_____

_____

_____

▲ These children are cleaning up litter.

Read a Photo

What has polluted this water?

_____

How can we save Earth's resources?

We need to save Earth's resources so we will have them for the future. We can reduce, reuse, and recycle to save Earth's resources.

To **reduce**, we cut back on how much we use something. To **reuse**, we use something again. To **recycle**, we make a new item out of an old item.

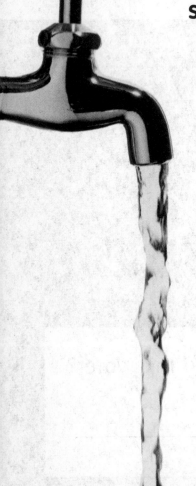

◀ Reduce how much water you use. Shut it off while you brush your teeth.

▲ Reuse something instead of throwing it away. This egg carton now holds paint.

▲ Recycle magazines into new paper. You can also recycle glass, metal, and many plastic items.

 Quick Check

11. Draw a picture of something you can reuse.
Write about how you will reuse it.

LOG ON e-**Review** Summaries and quizzes online at www.macmillanmh.com

Circle the correct answer.

1. When you make new items from old ones, you _____.

   reuse   recycle   reduce

2. A mix of tiny rocks and bits of plants and animals

   makes _____.

   soil   rocks   plants

3. Rocks, minerals, water, and air are examples of _____.

   plants   natural resources   pollution

4. When you cut back on how much you use something,

   you _____.

   reuse   recycle   reduce

5. Anything that makes air, water, or land dirty is _____.

   a natural resource   pollution   a mineral

Write what you learned.

_____

_____

CHAPTER 7

Observing Weather

How can we describe weather?

Vocabulary	
temperature how hot or cold something is	
precipitation water falling from clouds as rain, snow, or hail	
evaporate when water changes from a liquid to a gas	
condense when water changes from a gas to a liquid	
cumulus white puffy clouds	
cirrus high clouds made of ice	
stratus clouds that form into layers like sheets	

What is weather?

Weather is what it feels like outside. The weather can be hot or cold.

Temperature tells us how hot or cold something is. We can measure the temperature with a thermometer.

◀ This thermometer tells us that the air is warm.

◀ This thermometer tells us that the air is cool.

We can measure precipitation too.
Precipitation is water that falls to
Earth from clouds. Rain, snow, sleet,
and hail are all kinds of precipitation.

Rain can be measured with a rain gauge. ▼

▲ **Snow can be measured with a ruler.**

 Quick Check

I. What does a thermometer measure?

_____

_____

2. What are some kinds of precipitation?

_____

_____

What is wind?

Wind is air that moves. Differences in hot and cold air cause wind. Wind can be strong or light. We can measure wind with a wind sock. A wind sock points in the direction the wind is blowing.

wind sock

▲ A wind sock blows out straight if the wind is strong. It moves gently when the wind is light.

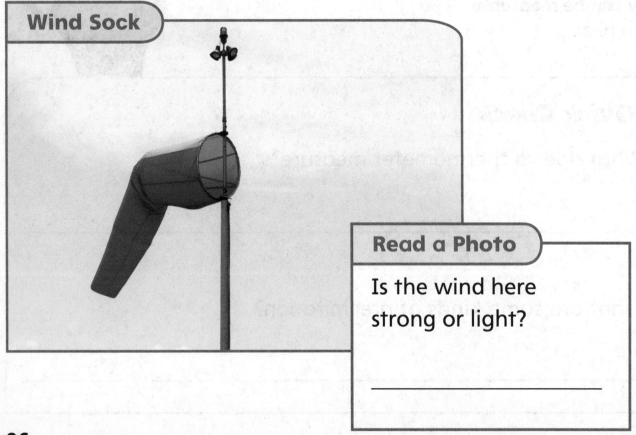

Wind Sock

Read a Photo

Is the wind here strong or light?

_____

Other tools measure the speed of wind. One tool has cups that catch the wind. It keeps track of how fast the cups spin. The cups spin quickly in a strong wind. They move slowly in a light wind.

This scientist is measuring a strong wind. ▶

 Quick Check

3. Draw a picture of a wind sock in strong wind.

How does water disappear?

Water we drink is a liquid. Liquid water can heat up. It can change into water vapor. Water vapor is a gas. When water changes from a liquid to a gas, it **evaporates**.

▼ **Water evaporates into water vapor over this pond.**

water vapor

liquid water

Water vapor rises in warm air. As the air cools, it can not hold the water vapor. The water vapor **condenses**, or turns back into a liquid. It makes tiny droplets that can form clouds.

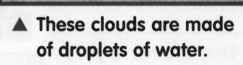

▲ These clouds are made of droplets of water.

✔ Quick Check

4. What happens when water evaporates?

_____

_____

5. What happens when water vapor condenses?

_____

_____

What is the water cycle?

Water evaporates from oceans, lakes, and rivers. The water vapor condenses and forms clouds. Water droplets in clouds fall back to Earth as precipitation. This is called the water cycle.

LOG ON **Science in Motion** Watch the water cycle at **www.macmillanmh.com**

The Water Cycle

The Sun heats the water. Water evaporates and rises into the air.

The water vapor cools. It condenses and forms clouds.

 **Quick Check**

6. Draw a picture that shows the water cycle.

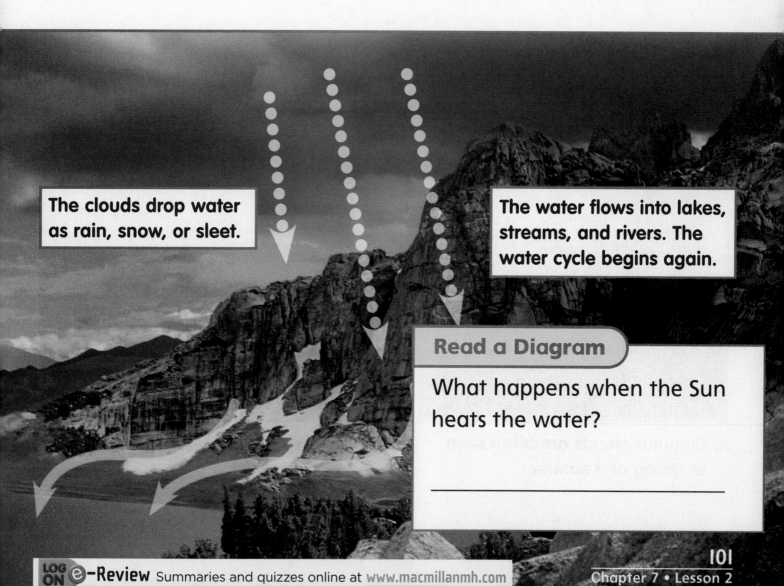

The clouds drop water as rain, snow, or sleet.

The water flows into lakes, streams, and rivers. The water cycle begins again.

Read a Diagram

What happens when the Sun heats the water?

_____

What are different kinds of clouds?

Not all clouds are the same. Each cloud tells us clues about the weather.

Cumulus clouds are white and puffy. They mean good weather.

Stratus clouds form large sheets low in the sky. They may mean a storm is coming.

▲ Cumulus clouds are often seen in spring and summer.

▲ Stratus clouds can be thick or thin.

Cirrus clouds are high, icy clouds. They mean the weather will change soon.

▲ Cirrus clouds are very high in the sky.

 Quick Check

Circle the answer.

7. If you see this kind of cloud, a storm may be coming.

stratus   cirrus   cumulus

8. Which kind of cloud means good weather?

cirrus   stratus   cumulus

How can we stay safe from weather?

Some types of weather can be dangerous. Thunderstorms can bring lightning and thunder. It is important to stay safe in very bad weather.

Lightning Safety

Stay out of water.

Do not stand under trees.

Stay indoors.

Read a Chart

Where should you stay during a lightning storm?

_____

Some storms grow very strong. They can turn into hurricanes. Some thunderstorms bring tornadoes.

◀ Tornadoes are spinning columns of air.

 Quick Check

Fill in the blanks.

9. A bad thunderstorm can bring a spinning column of

air called a _____.

10. Very strong storms can turn into

_____.

Draw a line from each word to its definition.

temperature

precipitation

evaporate

cumulus

cirrus

I. water falling from clouds as rain, snow, or hail

2. white puffy clouds

3. high clouds made of ice

4. a measurement of how hot or cold something is

5. when water changes from a liquid to a gas

Write what you learned.

_____

_____

CHAPTER 8

Earth and Space

What can we see in the night sky?

Vocabulary

rotation a turn or spin

axis a center, imaginary line that an object spins around

axis

orbit the path Earth takes around the Sun

phase the shape of the Moon we see from Earth

star an object in space made of hot gases that glow

planet a very big object that moves around the Sun

solar system the Sun, eight planets, and their moons

What causes day and night?

Earth spins all the time, but we can not feel it. The spinning of Earth is called **rotation**.

Earth's rotation causes day and night. It is day on the side of Earth facing the Sun. It is night on the other side.

Earth always rotates in the same direction. ▼

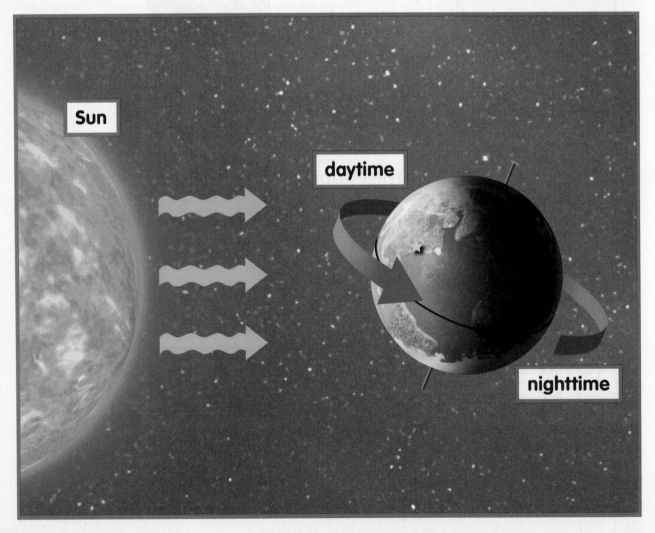

Sun

daytime

nighttime

Earth turns around an imaginary line called an **axis**. Earth turns all the way around in 24 hours, or one day.

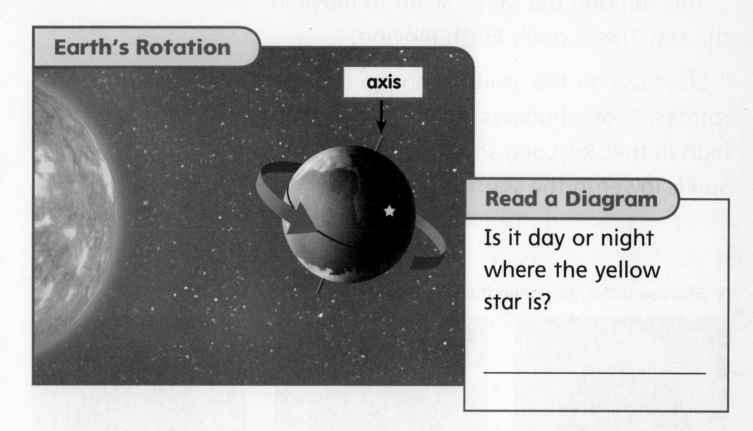

Earth's Rotation

axis

Read a Diagram

Is it day or night where the yellow star is?

_____

 Quick Check

Circle the answer.

1. What causes day and night?

   Earth's rotation   Sun's rotation   24 hours

2. How long does it take for Earth to spin all the way around?

   8 hours   16 hours   24 hours

Why do the Sun and Moon seem to move?

The Sun and the Moon seem to move in the sky. This is really Earth moving.

Shadows on the ground change as Earth rotates. Short shadows mean the Sun is high in the sky. Long shadows mean the Sun is lower in the sky.

▼ Shadows change during the day.

8:00 a.m.

12:00 noon

4:00 p.m.

▲ Each night the Moon seems to move across the sky.

 Quick Check

Write *true* if the sentence is true. Write *false* if the sentence is false.

3. Long shadows happen late in the day.

_____

4. The Moon makes different shadows during the day.

_____

What are the seasons like?

There are four seasons. Each one has its own kind of weather. Living things may change from season to season.

There are fewer daylight hours in the colder seasons. As it gets warmer, there are more daylight hours to play outside!

▲ The air can become cooler in fall. There are fewer hours of daylight.

▲ Winter is the coldest time of the year. Animals must keep warm.

Quick Check

Circle the correct answer.

5. There are _____ seasons.

three   four   six

6. The days are shorter in _____ seasons.

colder   warmer   summer

▲ In spring the air is warmer.
It can be rainy.

▲ Summer is the hottest season.
The days are long.

What causes the seasons?

Earth's **orbit** is the path it takes around the Sun. Earth takes about 365 days, or one year, to orbit the Sun.

spring

summer

fall

LOG ON *Science in Motion* Watch how Earth tilts at www.macmillanmh.com

Earth is tilted. As Earth moves around the Sun, the tilt of Earth causes the seasons. The part of Earth that tilts toward the Sun is warmer. The part of Earth that tilts away from the Sun is colder.

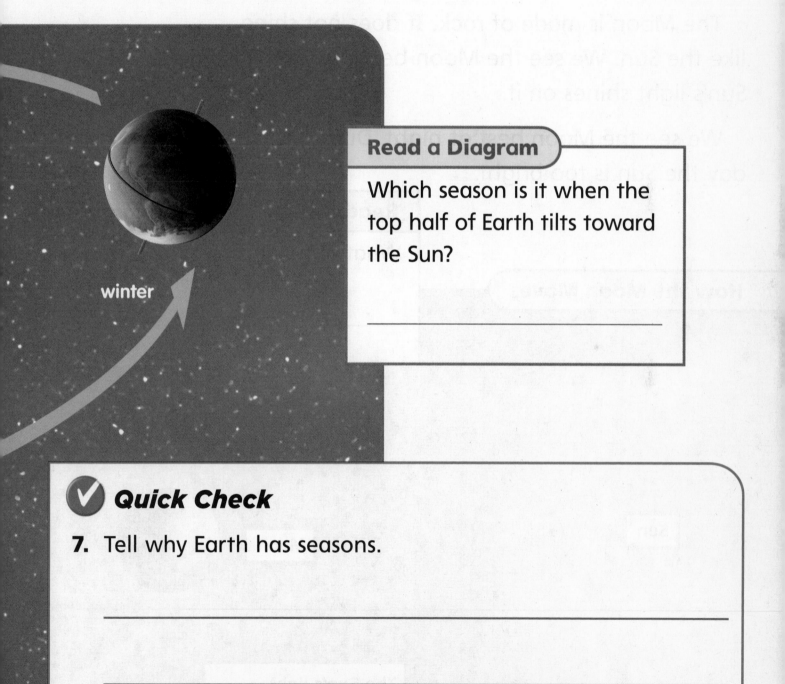

winter

Read a Diagram

Which season is it when the top half of Earth tilts toward the Sun?

_____

✓ **Quick Check**

7. Tell why Earth has seasons.

_____

_____

Why can we see the Moon from Earth?

The Moon is made of rock. It does not shine like the Sun. We see the Moon because the Sun's light shines on it.

We see the Moon best at night. During the day the Sun is too bright.

How the Moon Moves

Read a Diagram

What shines light on the Moon?

_____

Sun

Moon

The Sun's light shines on the Moon.

The Moon moves in an orbit around Earth. It takes about one month for the Moon to make one orbit. The Moon repeats this path again and again.

 Quick Check

8. Why do we see the Moon in the sky?

_____

9. Why do we see the Moon best at night?

_____

Earth

Why does the Moon seem to change shape?

Over time the Moon seems to change shape. One night it looks round. A few nights later it is thin.

The Moon's shape does not change. The amount of light shining on the Moon changes.

Each moon shape we see is called a **phase**. The phases of the Moon repeat in the same order each month.

Moon Phases

new moon

first quarter moon

❶ We can not see the Sun's light shining on the Moon.

❷ After a week the Moon is one quarter of the way through its orbit.

10. Draw what the Moon looks like about one week after a new moon.

full moon

3 The next week the Moon has moved. We can see all of the Moon's lit side.

last quarter moon

4 The Moon is three quarters of the way through its orbit by the third week.

What are stars?

A **star** is an object in space made of hot gases. The gases give off heat and light.

Some stars are bright. Stars can be different colors. Together some stars make patterns in the sky.

These lines show a pattern of stars called Orion the Hunter. ▼

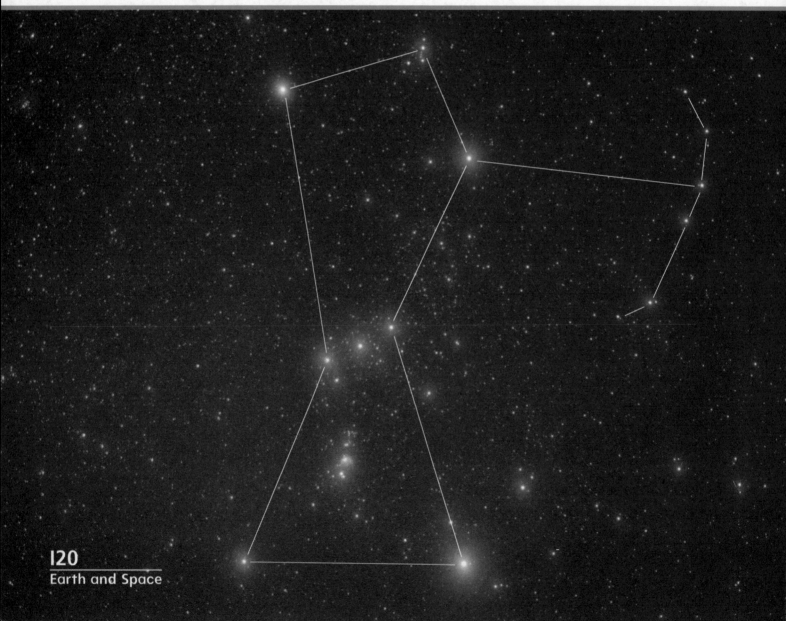

From Earth, stars look like tiny points of light. They are very, very far away. There is one star close to Earth. It is the Sun. It looks large because it is close to Earth.

The Sun lights up the sky during the day. We can not see other stars until night. ▶

✔ Quick Check

Circle the answer.

11. The _____ is a star that is close to Earth.

Sun   Moon   Orion

12. Stars look tiny because they are _____.

shining   far away   near

What goes around the Sun?

A **planet** is a huge object that moves around the Sun. We live on the planet Earth.

Our **solar system** is made of planets, moons, and the Sun. Solar means "of the Sun."

The Solar System

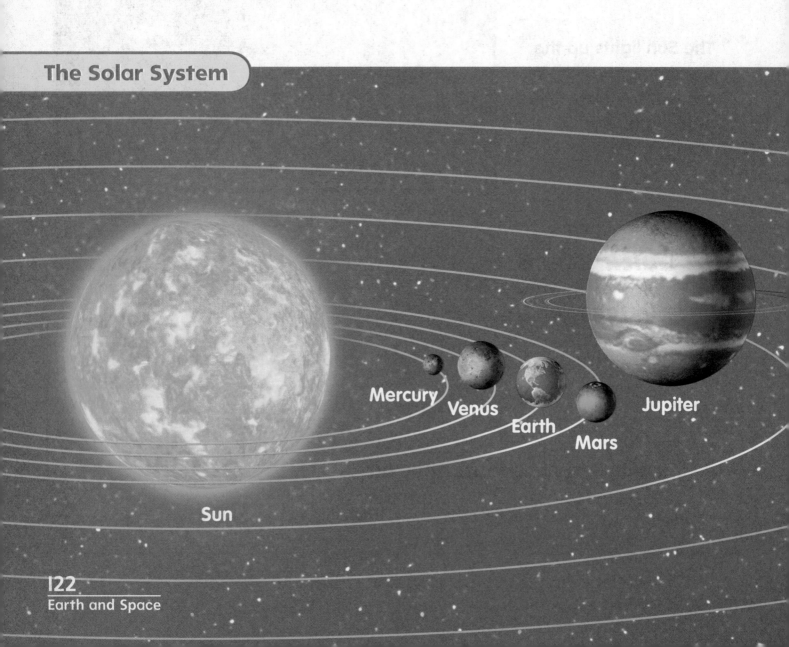

Mercury

Venus

Earth

Mars

Jupiter

Sun

The Sun is at the center of our solar system. It is the strongest and brightest part. Each planet moves around the Sun. Planets closest to the Sun take less time to make one trip around it.

 Quick Check

Fill in the blank.

13. Earth is a _____.

Read a Diagram

Which planet is the closest to the Sun?

_____

Saturn

Uranus

Neptune

What are the planets like?

Each planet in the solar system is different. There are four inner planets. They are closest to the Sun. There are four outer planets. They are farthest from the Sun.

Inner Planets

Mercury is the closest planet to the Sun.

Venus is the hottest planet.

Earth is where we live. It has one moon.

Mars has two moons. It is red and rocky.

Outer Planets

Jupiter is the largest planet. It has over 60 moons!

Quick Check

14. Which planet is red and rocky?

_____

15. Which planet has large rings around it?

_____

Saturn has large rings made of ice and rocks.

Uranus has rings. It has almost 30 moons.

Neptune is a blue planet with at least 13 moons.

Complete each sentence with a word from the box.

orbit	planet	rotation	solar system	star

1. The path Earth takes around the Sun is called its _____.

2. A big object that moves around the Sun is a(n) _____.

3. The Sun, the planets, and their moons make up our

   _____.

4. An object in space made of hot, glowing gases is a(n)

   _____.

5. The spinning of Earth is called _____.

Write what you learned.

_____

_____

CHAPTER 9

Looking at Matter

How can we describe matter?

Vocabulary

matter anything that takes up space and has mass

mass the amount of matter in an object

property how something looks, feels, smells, tastes, or sounds

solid matter that has a shape of its own

liquid matter that takes the shape of the container it is in

volume the amount of space something takes up

gas matter that spreads to fill the space it is in

What is matter?

Matter is anything that takes up space and has mass. **Mass** is the amount of matter in an object.

Matter can be made by people or found in nature. We use matter every day.

Using Matter

Read a Photo

What matter do you see?

Dog

Different objects have different masses. A truck has a lot of mass. A pencil has a little mass.

You can use a balance to measure and compare mass.

The big boot has more mass than the little boot.

Quick Check

Fill in the blanks.

1. The amount of matter in an object is its mass.

2. A tool used to measure the amount of mass in an object is a balance.

How can you describe matter?

We can describe matter by telling about its properties. A **property** is how matter looks, feels, smells, tastes, or sounds.

Matter can feel smooth or rough. Matter can be thick or thin. Matter can be soft or hard.

▲ **This mustard is thick and gooey.**

▲ **This skunk is white and black. It is also very smelly!**

There are many ways to talk about matter. Matter can be living or nonliving. Matter can be solid, liquid, or gas.

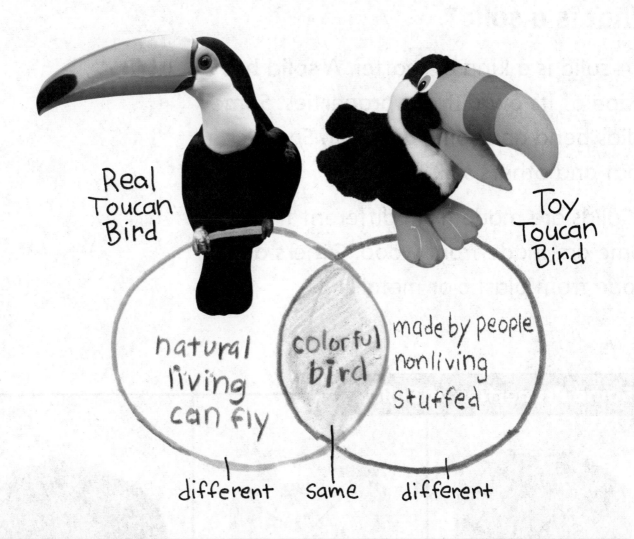

Real Toucan Bird

Toy Toucan Bird

natural living can fly

colorful bird

made by people nonliving stuffed

different   same   different

✓ Quick Check

3. Choose matter in your classroom. Describe its properties.

A pencil wirtes, smooth, and

orange yellow.

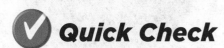

What is a solid?

A solid is a kind of matter. A **solid** has a shape of its own. It has properties. Some solids bend and some can tear. Some can float and others sink.

Solids are made from different things. Some are made from wood. Others are made from plastic or metal.

Some Properties of Solids

rock	glass	yarn
• **hard**	• **smooth**	• **soft**
• **spotted**	• **can break**	• **colorful**
• **rough**	• **clear**	• **long and thin**

Quick Check

4. Write a list of solids in your classroom.

Table, chair, crayon box, white bord,

clock, pencil.

5. Choose one solid from your list. List its properties.

A table is hard, wood, and smooth,

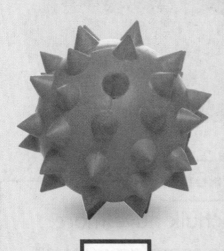

toy
- blue
- pointy
- plastic

sea sponge
- yellow
- squishy
- scratchy

clay
- sticky
- can bend
- firm

How can we measure solids?

We can measure solids in different ways. A balance tells how much mass something has. A ruler tells how long, wide, or tall something is. We can measure the same object in different ways.

Measuring Solids

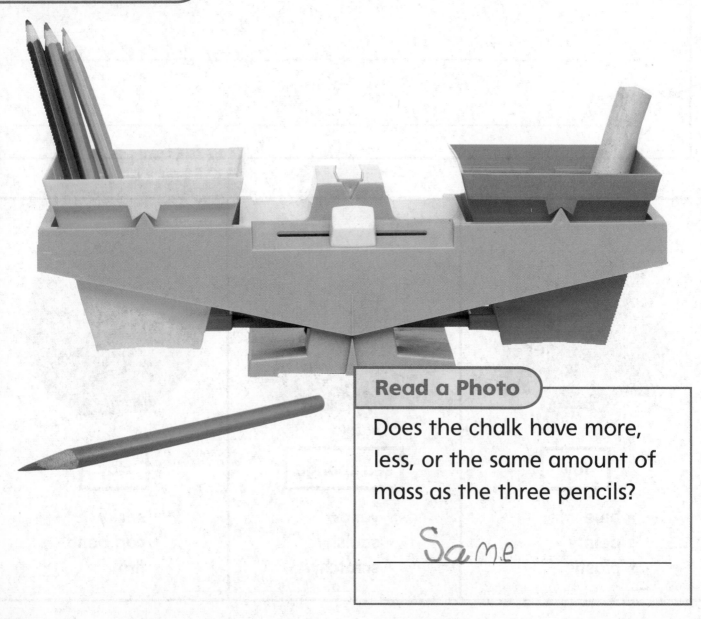

Read a Photo

Does the chalk have more, less, or the same amount of mass as the three pencils?

Same

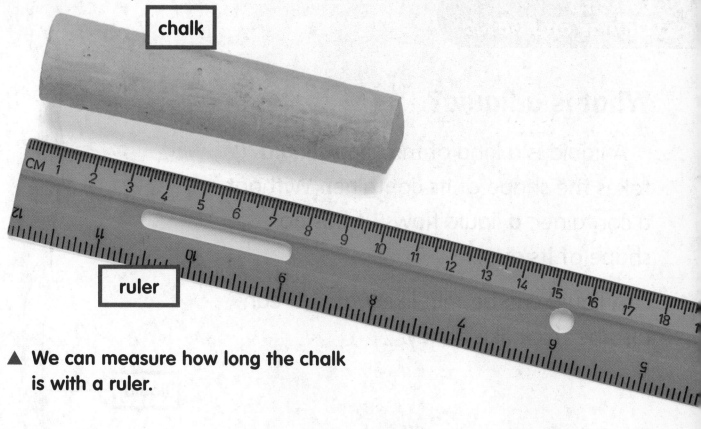

chalk

ruler

▲ We can measure how long the chalk is with a ruler.

✔ Quick Check

Fill in the blanks.

6. A ruler measures _length_.

7. A balance measures _mass_.

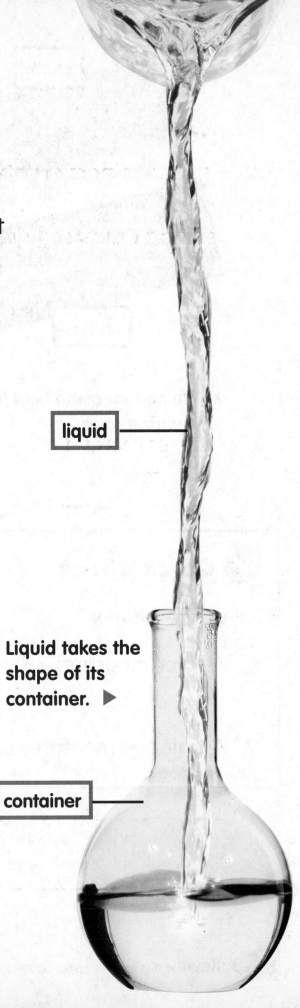

What is a liquid?

A liquid is a kind of matter. A **liquid** takes the shape of its container. Without a container, a liquid flows. It has no shape of its own.

A liquid can be thin like water. It can also be thick like honey.

This water flows over the cliff and takes the shape of the pond. ▼

liquid

Liquid takes the shape of its container. ▶

container

Volume is the amount of space something takes up. We can measure the volume of a liquid with a measuring cup.

These containers are the same size. They are holding different amounts of liquids.

20°C
Exml
100
90
80
70
60
50
40
30
20
10

A

20°C
Exml
100
90
80
70
60
50
40
30
20
10

B

Read a Photo

Does container A or container B have a greater volume of liquid?

A

 Quick Check

8. Make a list of liquids in your home.

Water, ketchp, pickle jucie, jucie, mustrd, soda, honey,

What is a gas?

A gas is a kind of matter. A **gas** spreads out to fill the space it is in. It does not have a shape of its own.

We can not see most gases, but they are everywhere. The air we breathe is made of many gases. We can feel air moving on a windy day.

Gases are inside these bubbles. ▼

▼ Gases are filling this toy.

All matter has mass. Even gases have mass. When we fill a balloon, the balloon gets more mass.

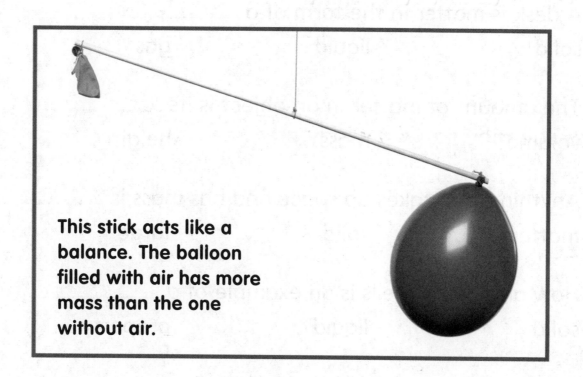

This stick acts like a balance. The balloon filled with air has more mass than the one without air.

Write *true* if the sentence is true. Write *false* if the sentence is false.

9. Gases have their own shape. _False_____

10. Gases have mass. _Ture_____

11. All matter has mass. _Ture_____

Circle the correct answer.

1. A desk is matter in the form of a ___solid___ .
   (solid)   liquid   gas

2. The amount of matter in an object is its ___mass___ .
   (volume)   (mass)   height

3. Anything that takes up space and has mass is ___matter___ .
   (matter)   solid   large

4. How an object smells is an example of a ___property___ .
   solid   liquid   (property)

5. Matter that has no shape of its own is a ___liquid___ .
   solid   (liquid)   property

6. Matter that spreads to fill the space it is in is a ___gas___ .
   (gas)   solid   mass

Write what you learned.

_____

_____

CHAPTER 10

Changes in Matter

How can matter change?

Vocabulary	
physical change a change in the size or shape of matter	
chemical change when the properties of matter change	
evaporate to change from a liquid to a gas	
condense to change from a gas to a liquid	
mixture two or more things put together	
solution a kind of mixture that is hard to take apart	

What are physical changes?

Matter can change in different ways. A **physical change** happens when the shape or size of matter changes. Folding and bending matter changes its shape. Cutting and tearing matter changes its size.

Folding does not change the properties of the paper. It is still paper. ▶

◀ **Folding does not change the mass of the paper.**

A physical change can happen when the temperature of matter changes. On a cold day, water turns into ice. Wetting and drying are physical changes too.

wet mud

dry mud

 Wet mud feels squishy.

▲ Dry mud is hard and light in color.

✔ Quick Check

Draw an object in the first box. Then draw what it would look like after a physical change.

I.

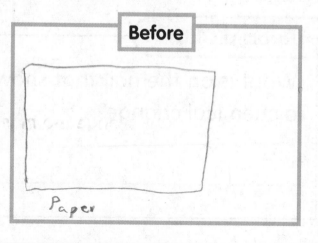

Before

Paper

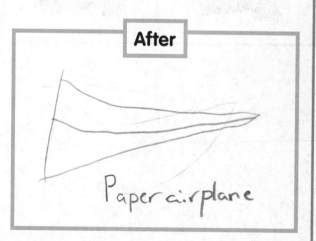

After

Paper airplane

What are chemical changes?

A **chemical change** happens when the properties of matter change. It is not easy to change matter back after this kind of change. It becomes a new kind of matter.

A chemical change happens when something burns or rusts. Cooking causes this kind of change too.

Chemical Changes		
Before	**After**	**Cause**
		Heat caused the matchstick to burn. Its properties have changed.
		Water and air caused this metal nail to rust.
		Water and air do not change the properties of plastic.

Read a Chart

What is on the nail that shows a chemical change?

rust

All matter does not change in the same way. Seeing light and feeling heat can mean a chemical change is happening.

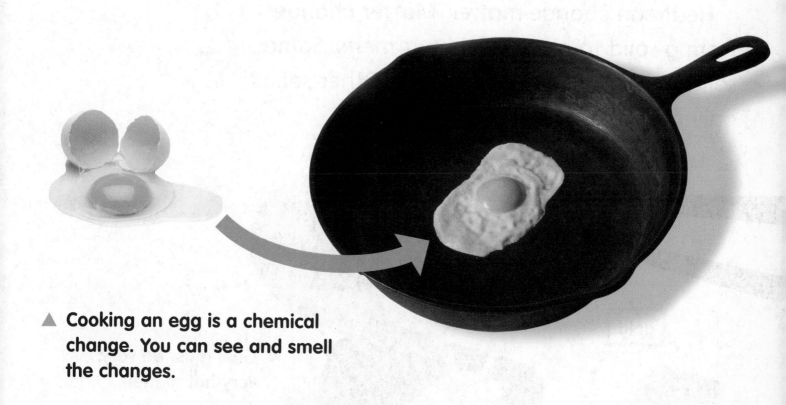

▲ Cooking an egg is a chemical change. You can see and smell the changes.

✔ Quick Check

Write *true* if the sentence is true. Write *false* if the sentence is false.

2. All matter changes in the same way. _false_

3. Burning and rusting are chemical changes. _True_

How can heating change matter?

Heat can change matter. Matter changes from a solid to a liquid when it melts. Some solids melt at low temperatures. Other solids only melt when they are very hot.

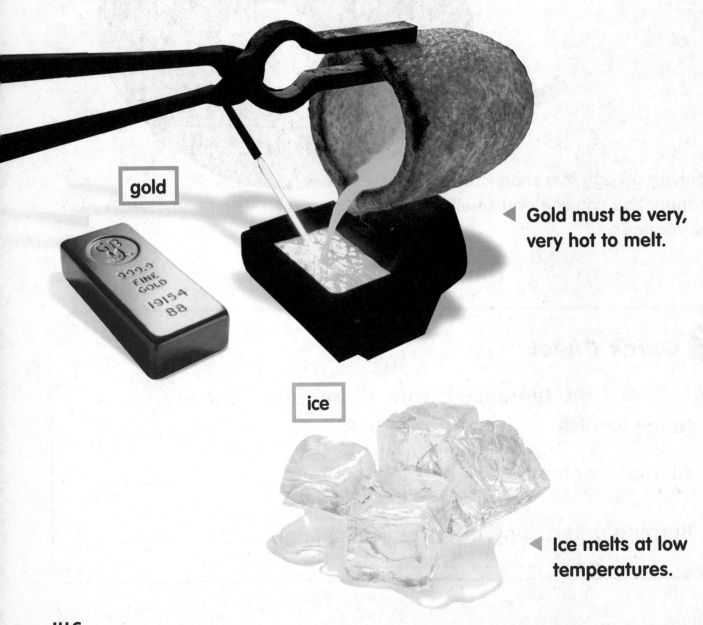

gold

◀ Gold must be very, very hot to melt.

ice

◀ Ice melts at low temperatures.

Matter changes from a liquid to a gas when it **evaporates**. Water begins to boil when it is heated. Bubbles show that the water is turning into water vapor. Water vapor is a gas.

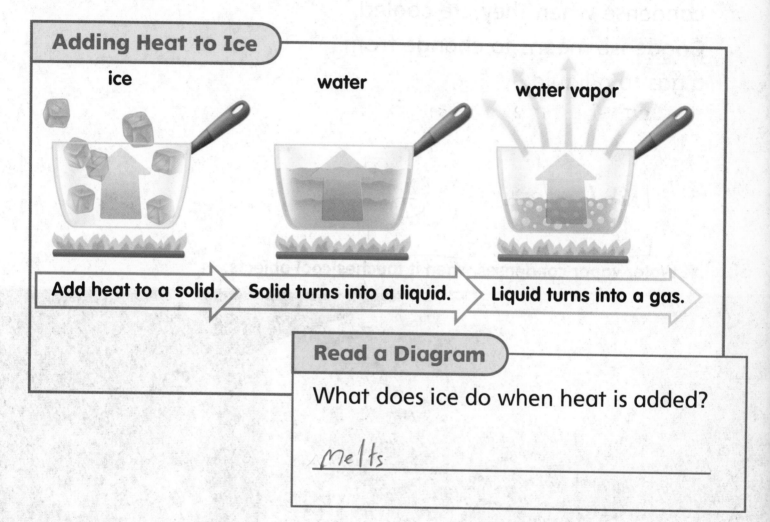

Adding Heat to Ice

ice   water   water vapor

Add heat to a solid. ▷ Solid turns into a liquid. ▷ Liquid turns into a gas. ▷

Read a Diagram

What does ice do when heat is added?

melts

✓ **Quick Check**

4. How can ice change into water vapor?

When it gets really hot and Boils.

How can cooling change matter?

Cooling can change matter. Cooling means heat is taken away. Gases condense when they are cooled. **Condense** means to change from a gas to a liquid.

▼ Water vapor condenses when it touches cool objects.

When liquids cool, they can freeze or turn into a solid. Wax and some liquids are solid at room temperature. Water and other liquids need to be very cold to freeze.

**The hot wax is liquid.
The cool wax is solid.** ▶

✔ *Quick Check*

5. What happens when a liquid is cooled?

They freeze or turns soild.

6. What happens when a gas is cooled?

Condeses gass into liusoo

What are mixtures?

A **mixture** is two or more things put together. Any combination of solids, liquids, or gases can be a mixture. Salt stirred into water is a mixture. Bits of clay put together make a mixture too.

▼ This bowl contains a mixture of flour, water, and newspaper.

Some mixtures are easy to take apart. The parts of the mixture are easy to see. The things in the mixture do not change.

This cup holds a mixture. The parts can be taken apart. ▶

 Quick Check

7. Draw a picture of a mixture that can be taken apart.

Which mixtures stay mixed?

Sometimes it is hard to take mixtures apart. It may not be easy to change the parts of the mixture back.

Making a Smoothie

before

after

Read a Photo

Which mixture can be taken apart?

The Before

A **solution** is a mixture that is hard to take apart. Sugar and water make a solution. The sugar **dissolves**, or stays evenly mixed in the water.

Sand and water do not make a solution. The sand does not dissolve. It sinks to the bottom of a glass.

This solid dissolves in this liquid to make a solution. ▶

 Quick Check

8. What is a solution?

A solution is a mixture that is hard to take apart.

How can you take mixtures apart?

There are different ways to take apart a mixture. Filters can trap solids but let liquids go through. Magnets can separate mixtures with iron and other metals.

A filter can take apart a mixture of sand and water. ▼

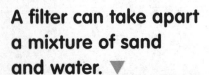

magnet

▲ **A magnet can take apart a mixture of sand and iron bits.**

filter

Some mixtures are very hard to take apart. Salt water is hard to take apart. The water must evaporate. This leaves the salt behind.

▲ **Water from the ocean can evaporate. The salt is left behind.**

✓ **Quick Check**

9. What are different ways a mixture can be taken apart?

With your hands, filter and

magnet.

Use the words in the box to complete each sentence.

chemical change	mixture
condenses	physical change

1. Two or more things put together are called a _____.

2. A change in the size or shape of matter is called a

   _____.

3. When a gas changes into a liquid, it _____.

4. A kind of change where matter turns into different matter is a

   _____.

Write what you learned.

_____

_____

CHAPTER 11

How Things Move

How do things move?

Vocabulary	
position the place where something is	
force a push or pull on something	
gravity a force that pulls objects toward Earth	
friction a force that slows down moving things	
attract to pull toward something	
repel to push away or apart	

What are position and motion?

Position is the place where something is. Words such as above or below tell where things are. Other words such as next to, in, under, and on also tell about the position of an object.

▼ The dog is on the stool. The cat is in the wagon. The toy car is under the stool.

Motion is a change in the position of an object. Objects can move up, down, sideways, or zigzag. We can tell about an object's motion by telling how its position changed.

This dog is jumping up. ▶

◀ **This dog is running in a zigzag.**

 Quick Check

I. Draw a picture with at least three objects. Write a sentence telling about one object's position.

What is speed?

Speed is how far something moves in a certain amount of time. Some things move fast. Some things move slowly.

▲ A snail's speed is slow.

A cheetah's speed is fast. ▶

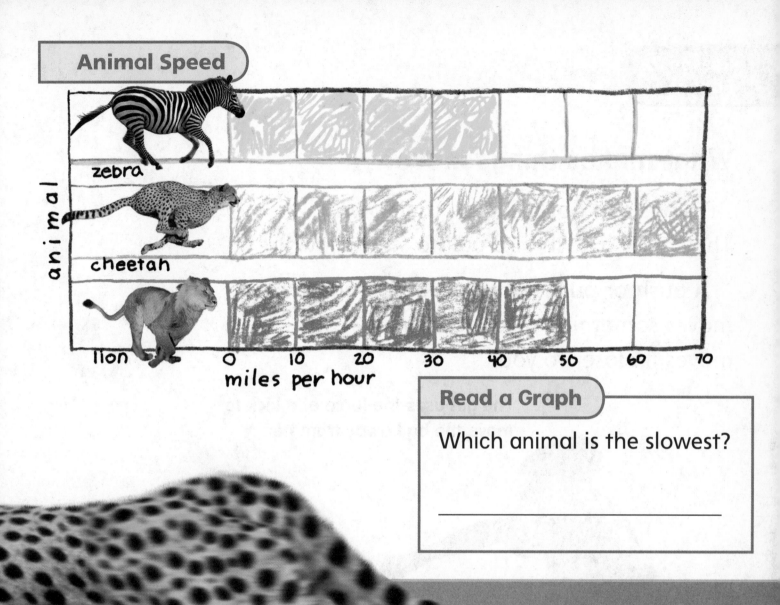

Animal Speed

animal

zebra

cheetah

lion

0   10   20   30   40   50   60   70

miles per hour

Read a Graph

Which animal is the slowest?

_____

✓ Quick Check

Fill in the chart. What moves slowly? What moves fast?

2.

Moves Slowly	Moves Fast

What makes things move?

Things can not move on their own.
They need a push or a pull to start moving.

A push or pull is called a **force**. A push
moves something away from you. A pull
moves it closer to you.

**The girl uses the force of a kick to
move the ball away from her.** ▼

We push and pull things all the time. A kick is a push. Opening a drawer is a pull. Different objects need different amounts of force to move.

A big push gets this cart moving. ▶

 Quick Check

Fill in the blanks.

3. A push or a pull is called a _____.

4. Opening a drawer is a kind of _____.

5. A kick is a kind of _____.

The children are pulling in different directions. ▼

What are some forces?

Gravity is a force that pulls things toward Earth. Gravity is what keeps you on the ground.

Gravity pulls down on things through solids, liquids, and gases. The amount of force that pulls something down toward Earth is called its weight.

Gravity will pull this dog back to the ground. ▼

Friction is a force that slows down moving things. There is friction when two things rub together.

Rough surfaces usually cause more friction than smooth ones. It is harder to move something over a rough surface than over a smooth one.

rubber stopper

▲ A skater drags her rubber stopper. The friction slows her down.

✔ *Quick Check*

Circle the answer.

6. A _____ surface causes more friction.

smooth   shiny   rough

7. We stay on the ground because of _____.

friction   gravity   surfaces

How can forces change motion?

Forces can make things start moving and speed up. Forces can make things slow down and stop. Forces can make things change direction too.

How a Ball Changes Direction

1. The pitcher uses a force to move the ball toward the batter.

2. The batter uses a push to hit the ball. The ball changes direction.

LOG ON *Science in Motion* Watch forces work at www.macmillanmh.com

3. The player catches the ball. He stops its motion.

Read a Diagram

If the player threw the ball back to the pitcher, would it be a push or a pull?

_____

✓ Quick Check

Circle all correct answers.

8. Forces can make things _____.

   speed up   slow down   change direction

9. When the player catches the ball, its motion _____.

   stops   speeds up   starts

What are levers and ramps?

A **simple machine** is a tool that makes moving an object easier. It makes the force of a push or a pull stronger.

A lever is a simple machine with a bar that moves on a point. Shovels, seesaws, and hammers are levers.

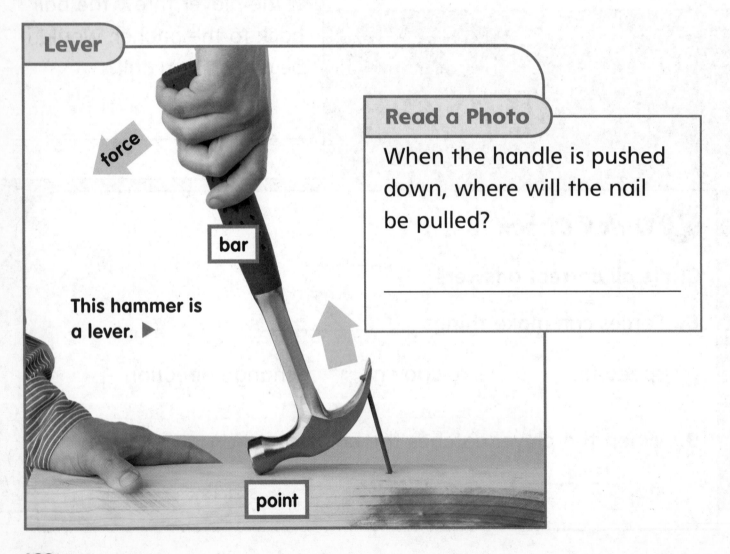

Lever

force

bar

This hammer is a lever. ▶

point

Read a Photo

When the handle is pushed down, where will the nail be pulled?

_____

▲ This is a ramp. A heavy box can be pushed up a ramp.

A ramp is a simple machine that helps us move things to a higher place. A ramp is a surface with one end raised up. It is easier to push an object up a ramp than it is to lift it.

 Quick Check

10. How we can use simple machines to do work?

_____

_____

What are other simple machines?

A wheel is a kind of simple machine. A wheel works with a bar that is attached to its center. When the wheel turns, this bar turns too.

A bar connects two wheels. This truck has two sets of wheels. ▼

wheel

bar

A pulley is a simple machine with a rope that moves around a wheel. A pulley can help lift an object.

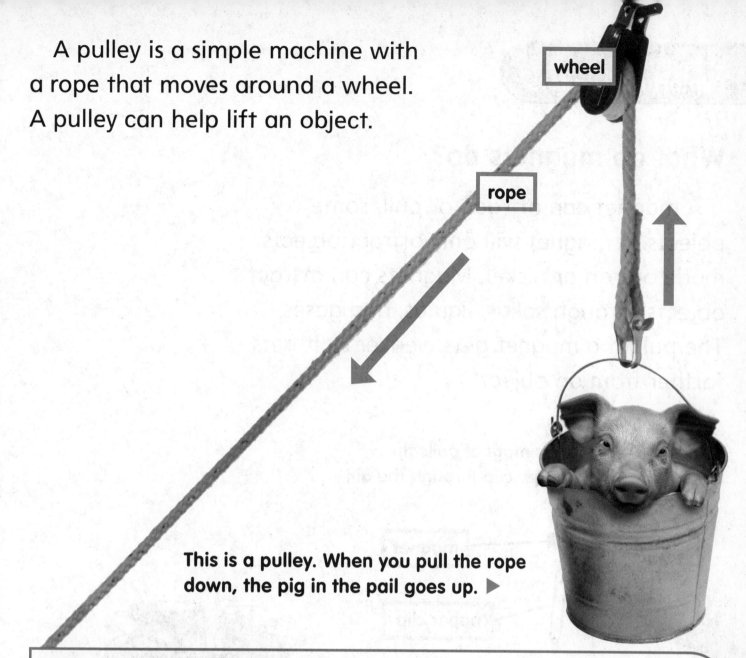

wheel

rope

This is a pulley. When you pull the rope down, the pig in the pail goes up. ▶

✓ Quick Check

Fill in the blanks.

11. A _____ connected to a bar is a kind of simple machine.

12. A pulley helps us _____.

What do magnets do?

A magnet can **attract**, or pull, some objects. A magnet will only attract objects made of iron or nickel. Magnets can attract objects through solids, liquids, and gases. The pull of a magnet gets weaker as it gets farther from an object.

▼ **This magnet pulls the paper clip through the air.**

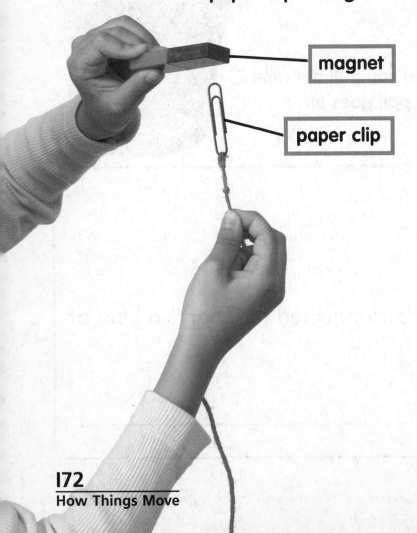

magnet

paper clip

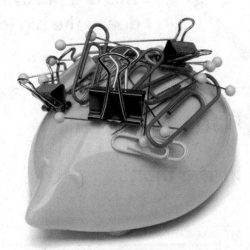

▲ **This magnet attracts many objects.**

There are some objects magnets can not attract. Plastic, wood, and some metals can not be pulled by a magnet.

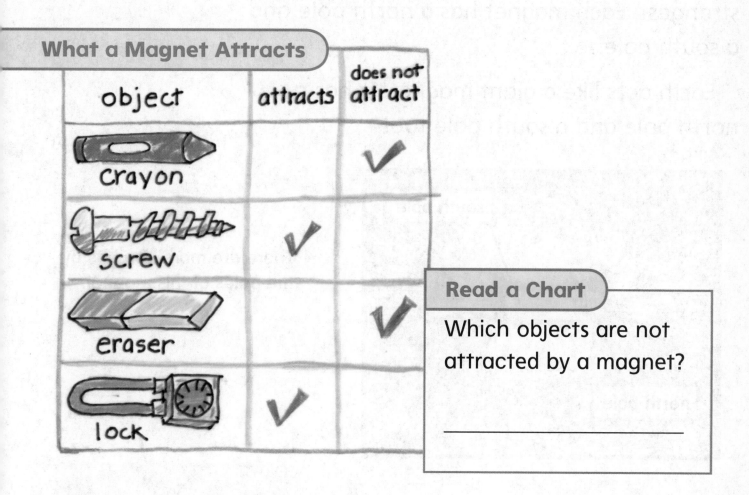

What a Magnet Attracts

object	attracts	does not attract
crayon		✔
screw	✔	
eraser		✔
lock	✔	

Read a Chart

Which objects are not attracted by a magnet?

_____

✔ *Quick Check*

Write *true* if the sentence is true. Write *false* if the sentence is false.

13. Magnets pull objects with iron in them. _____

14. Magnets can pull plastic. _____

What are poles?

Each end of a magnet is called a **pole**. The poles are where the pull of the magnet is the strongest. Each magnet has a north pole and a south pole.

Earth acts like a giant magnet. It has a north pole and a south pole too!

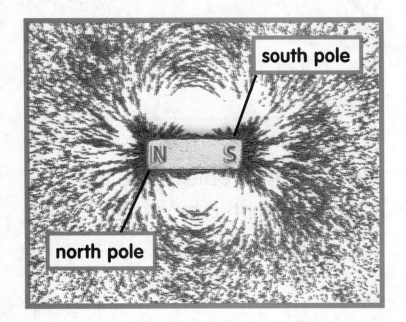

south pole

north pole

◀ There are more iron bits by the poles of this magnet.

Earth has magnetic force around its poles. ▶

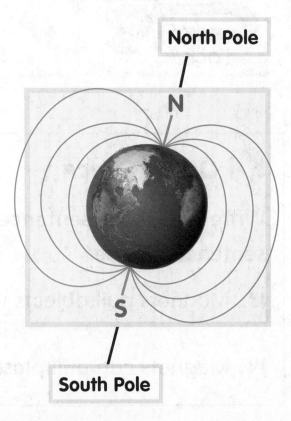

North Pole

N

S

South Pole

A north pole and a south pole pull toward each other.

◀ The north and south poles attract each other.

Two of the same poles placed together will **repel**, or push away.

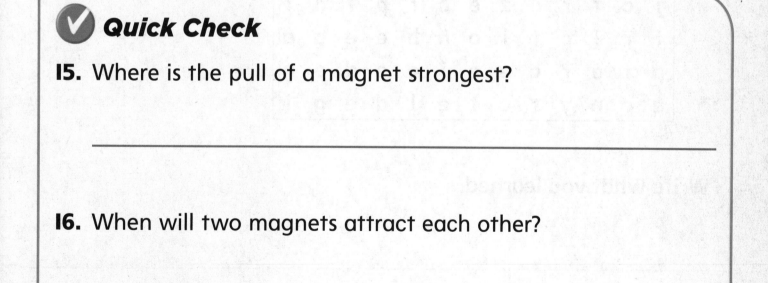

◀ Two south poles repel each other.

✔ Quick Check

15. Where is the pull of a magnet strongest?

_____

_____

16. When will two magnets attract each other?

_____

_____

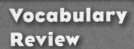
Use the clues to find the words. They go across
and down.

1. a change in the position of something

2. the push or pull on an object

3. a force that slows down moving things

4. how far something moves in a certain
amount of time

5. the two ends of a magnet

```
s x m a n e q z c r i y e
a p o l e s r a m s y m g
j c t r o z e c r p l n f
f r i c t i o n b e e a a
p f o r c e e i e e d e n
q c n y r c t e l d e g i
```

Write what you learned.

_____

_____

CHAPTER 12

Using Energy

The Big Idea

How do we use energy?

Vocabulary

heat a kind of energy that warms things	
fuel something that gives off heat when it burns	
sound a kind of energy that we can hear	
vibrate to move back and forth quickly	
light a kind of energy that lets us see	
circuit a path that electricity flows in	

What is heat?

Energy makes matter move or change.
Heat is a kind of energy that warms things.
Heat can change a solid to a liquid. Heat can
change a liquid to a gas. Most heat on Earth
comes from the Sun.

▼ **The Sun warms land, water, and air.**

Heat comes from other things too. **Fuel** is something that gives off heat when it burns. Gas, oil, wood, and coal are fuel.

Heat can also come from motion. When two things rub together, they heat up.

▲ **People use fuel to cook food.**

◄ **This motion makes heat.**

✔ Quick Check

Circle the answer.

1. People use fuel for _____.

heat   ice   lifting

2. Gas and wood are two kinds of _____.

motion   fuel   heat

What is temperature?

Temperature is how hot or cold something is. We use a thermometer to measure temperature. Some thermometers have liquid inside. The liquid goes up when the temperature gets warmer. It goes down when the temperature gets cooler.

Temperature

Read a Photo

Is it hotter here during the day or night?

_____

We can measure the temperature of air, water, and soil. We can even measure the temperature of our bodies!

Our temperature can go up when we are sick. ▶

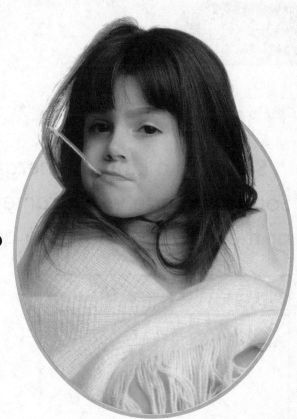

 Quick Check

3. What do we use to measure temperature?

_____

_____

4. What are some temperatures we can measure?

_____

_____

What makes sound?

Sound is a kind of energy you hear.
Sound is made when something **vibrates**,
or moves back and forth quickly.

How We Hear Sound

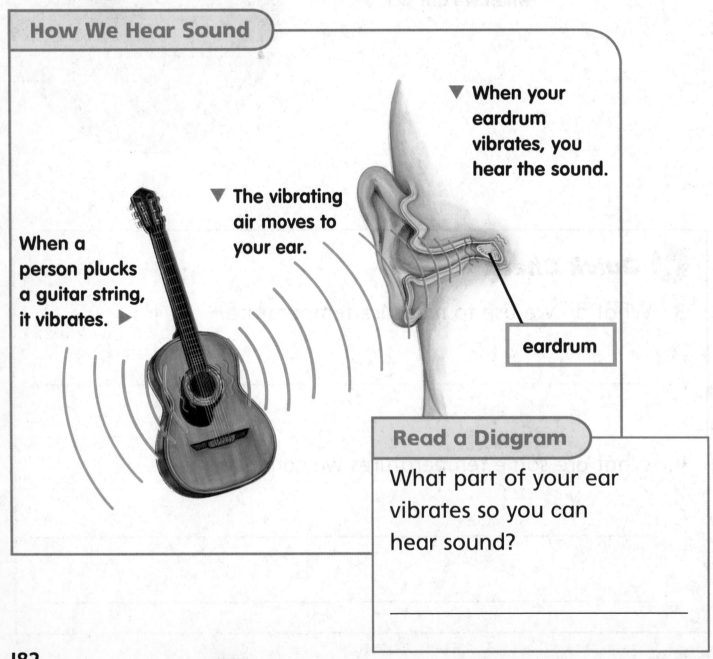

▼ **When your eardrum vibrates, you hear the sound.**

▼ **The vibrating air moves to your ear.**

When a person plucks a guitar string, it vibrates. ▶

eardrum

Read a Diagram

What part of your ear vibrates so you can hear sound?

_____

LOG ON *Science in Motion* Watch how sound travels at www.macmillanmh.com

First an object vibrates. This causes the air around it to vibrate. Then the vibrating air reaches your ears.

The eardrum begins to vibrate. It sends messages to your brain. Your brain tells you what sound you heard.

▲ The bells on the alarm clock ring. They vibrate to make sound.

 Quick Check

5. How does a guitar make sound?

_____

_____

How are sounds different?

There are different types of sounds. Sounds can be soft or loud. A soft sound has less energy than a loud sound.

A cat's meow is a soft sound. ▶

▼ The roar of a lion is a loud sound.

Pitch is how high or low a sound is. A sound with a high pitch vibrates quickly. A sound with a low pitch vibrates slowly.

A short string makes a sound with high pitch. ▶

harp

A long string makes a sound with low pitch ▶

 Quick Check

Write *true* if the sentence is true. Write *false* if the sentence is false.

6. A sound with a low pitch vibrates slowly.

_____

7. Pitch is how loud or soft a sound is.

_____

What do sounds move through?

Sounds can move through solids, liquids, and gases. Sounds move through a door or a window. We can hear sounds under water. Most sounds we hear travel through air.

▼ **Sound moves under water. Dolphins can hear each other's sounds.**

The closer we are to a sound, the louder it is. The farther away we are from a sound, the softer it is.

▲ The fire engine's siren will be louder as it gets closer.

> ## ✓ Quick Check
>
> **Fill in the blanks.**
>
> 8. Most sounds we hear travel through _____.
>
> 9. Sounds that are near will sound _____.

What is light?

Light is a kind of energy that lets us see things. You see because light will **reflect,** or bounce, off things. The reflected light goes into your eyes. Then you see the objects.

▼ **Most light on Earth comes from the Sun.**

Some objects let light through. Some do not. Glass can let light through. A book can block light and make a shadow. A shadow is a dark area where light does not reach.

Shadows

Read a Photo

What happens when the puppet blocks light?

_____

✓ Quick Check

10. Write a list of objects that light can not pass through.

_____

_____

_____

How do we see color?

Light is a mix of all colors. Light splits into different colors when it bends. We see the colors of the rainbow.

A prism can bend light. Raindrops can bend light too.

Raindrops can bend light into a rainbow. ▼

▲ **White light is a mix of many colors.**

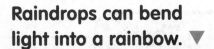

A filter is a tool that lets only certain colors of light pass through. A red filter blocks all colors but red. A green filter blocks all colors but green.

▲ Colored glass is a kind of filter.

✓ Quick Check

Circle all correct answers.

11. Light is a mix of _____.

    all colors   red and blue   three colors

12. A _____ can bend light.

    filter   prism   raindrop

What is current electricity?

Current electricity is a kind of energy
that moves along a path. The path is
called a **circuit**. Electricity can not move
if the circuit is not connected.

Circuit

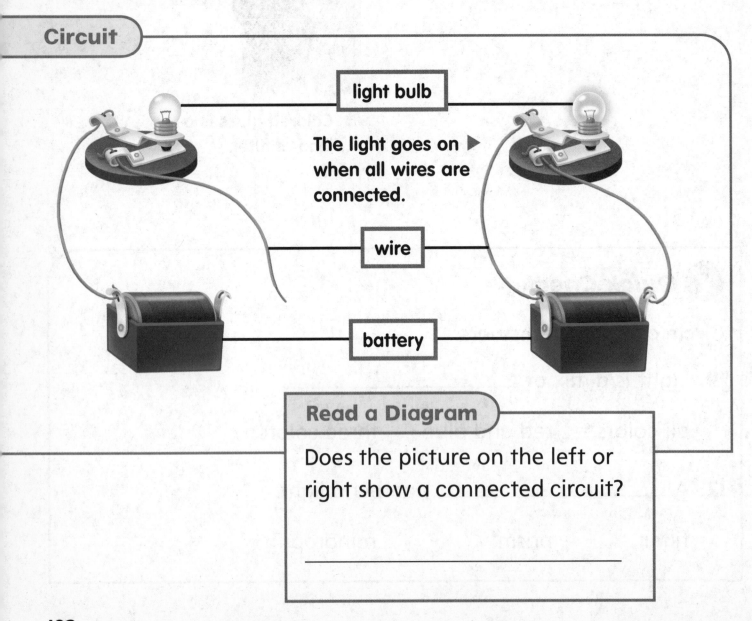

light bulb

The light goes on ▶
when all wires are
connected.

wire

battery

Read a Diagram

Does the picture on the left or
right show a connected circuit?

_____

Current electricity can be changed into heat, light, or sound energy. It can come from batteries or from outlets in the wall of a home.

Power plants change other kinds of energy into electricity. The electricity runs through wires to your house. We plug things in to complete a circuit and have power.

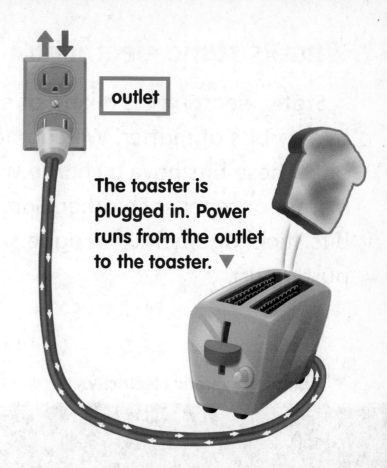

outlet

The toaster is plugged in. Power runs from the outlet to the toaster. ▼

✔️ *Quick Check*

13. How do we get electricity into our homes?

_____

_____

14. Write how you use electricity in your home.

_____

_____

What is static electricity?

Static electricity is a kind of energy made by tiny bits of matter. We can not see these bits. These bits have a charge when they attract or repel each other. Some of the bits stick together like magnets. Others push apart.

▼ **Lightning is static electricity.**

You can make static electricity by rubbing some things together. The objects become charged. Sometimes you can see or hear a static charge.

▲ The cat's fur is attracted to the charged balloon. The fur sticks up.

 **Quick Check**

15. What is static electricity?

_____

_____

16. What could make two socks stick together after they come out of the dryer?

_____

_____

Write the letter of the correct definition next to each word.

1. _____ heat

    a. to move back and forth quickly

2. _____ sound

    b. a kind of energy you hear

3. _____ vibrate

    c. a path that electricity flows in

4. _____ light

    d. a kind of energy that helps us see things

5. _____ circuit

    e. a kind of energy that makes objects warmer

Write what you learned.

_____

_____

Glossary

A

amphibian Animal that lives part of its life in water and part on land. (page 19) A salamander is an amphibian.

Arctic A very cold place near the North Pole. (page 54) Animals in the Arctic have layers of fat to keep them warm.

attract To pull toward something. (page 172) A magnet can attract some objects.

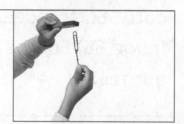

axis A center, imaginary line that an object spins around. (page 109) Earth spins on its axis.

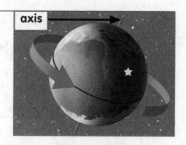

C

chemical change When matter changes into different matter. (page 144) Cooking an egg makes a chemical change.

circuit A path that electricity flows in. (page 192) A bulb will light when it is part of a connected circuit.

cirrus High clouds made of ice. (page 103) The wind blows cirrus clouds into thin pieces.

condense To change from a gas to a liquid. (pages 99, 148) Water vapor can condense on a cold bottle.

core Earth's deepest and hottest layer. (page 66) Earth's core is many miles below our feet.

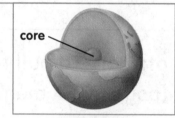

crust Earth's outer layer. (page 66) We live on Earth's crust.

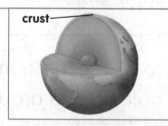

cumulus White, puffy clouds. (page 102) Small cumulus clouds are in the sky in good weather.

current electricity A kind of energy that moves in a path. (page 192) You use current electricity when you use a toaster.

desert A dry habitat that gets very little rain. (page 52) A desert can be very hot and dry.

dissolve To stay evenly mixed in a liquid and make a solution. (page 153) Drink mix will dissolve when it is mixed with water.

E

earthquake A shake in Earth's crust. (page 74) An earthquake damaged this road.

endangered When few of one kind of animal are left. (page 43) These tigers are endangered.

evaporate To change from a liquid to a gas. (pages 98, 147) Water can evaporate from lakes and ponds.

extinct When a living thing dies out and no more of its kind live on Earth. (page 45) Dinosaurs are extinct.

flood Water that moves over land and does not soak into the ground. (page 75) The man is walking in a flood.

flower Plant part that makes seeds or fruit. (page 6) Some flowers can grow into fruit.

food chain A model that shows the order in which living things get the food they need. (page 36) A food chain begins with the Sun.

food web Two or more food chains that are connected. (page 39) This picture shows a desert food web.

force A push or pull on something. (page 162) You use force to kick a ball.

fossil What is left of a living thing from the past. (page 44) This is a fish fossil.

friction A force that slows down moving things. (page 165) A skate makes friction when the wheels rub against the ground.

fuel Something that gives off heat when it burns. (page 179) Wood is fuel for fire.

 G

gas Matter that spreads to fill the space it is in. (page 138) The tube is filled with gas.

gravity A force that pulls objects toward Earth. (page 164) Gravity pulls this ball to the ground.

H

habitat A place where plants and animals live. (page 32) A habitat can be wet, dry, hot, or cold.

heat A kind of energy that warms things. (page 178) The Sun gives us heat.

I

insect Animal with six legs and an outer shell. (page 20) An ant is an insect.

L

landform One of the different shapes of Earth's land. (page 62) This landform is called a valley.

landslide When rocks and soil slide from higher ground to lower ground. (page 75) Things can be damaged in a landslide.

larva A stage in the life cycle of some animals after they hatch from an egg. (page 24) A caterpillar is a larva.

life cycle Steps that show how a living thing grows, changes, and makes new living things. (page 10) This picture shows parts of a life cycle.

light A kind of energy that lets us see. (page 188) We get light from the Sun.

liquid Matter that takes the shape of the container it is in. (page 136) Water is a liquid.

litter Garbage that is not thrown away. (page 88) People can clean up litter.

M

mammal Animal with hair or fur that feeds milk to its babies. (page 18) A lion is a mammal.

mantle A very hot layer below Earth's crust. (page 66) The mantle is too hot for living things.

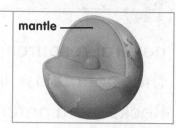

mantle

mass The amount of matter in an object. (page 128) The larger boot has more mass.

matter Anything that takes up space and has mass. (page 128) Everything around us is made of matter.

minerals Bits of rock and soil that help living things grow. (pages 4, 80) Plants use minerals in the ground to grow.

mixture Two or more things put together. (page 150) This snack food is a mixture.

motion A change in the position of something. (page 159) The dog is in motion.

N

natural resource Something from Earth that people use in daily life. (page 78) Rocks are a natural resource.

O

ocean A large body of salty water. (page 56) Dolphins live in the ocean.

orbit The path Earth takes around the Sun. (page 114) The Sun is at the center of Earth's orbit.

oxygen A gas found in the air we breathe. (page 5) Living things need oxygen.

 P

phase The shape of the Moon we see from Earth. (page 118) The phase of the Moon changes each night.

physical change A change in the size or shape of matter. (page 142) Folding paper is a physical change.

pitch How high or low a sound is. (page 185) Short strings make a high pitch.

high pitch

planet A very big object that moves around the Sun. (page 122) Mercury is the planet closest to the Sun.

poles Parts of a magnet where its pull is strongest. (page 174) A magnet has a north pole and a south pole.

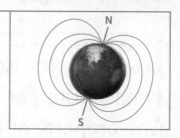

N

S

pollen A sticky powder inside flowers. (page 6) Pollen can move from flower to flower.

pollution Anything that makes air, land, or water dirty. (page 88) Garbage is one kind of pollution.

pond A small body of fresh water. (page 58) A pond is a home to plants and animals.

position The place where something is. (page 158) The position of the dog is above the cat.

precipitation Water falling from clouds as rain, snow, or hail. (page 95) Rain is one kind of precipitation.

property How something looks, feels, smells, tastes, or sounds. (page 130) One property of this toy is that it is soft.

pupa A stage in an insect's life cycle when a larva makes a hard case around itself. (page 24) This pupa is hanging from a branch.

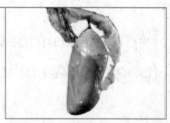

 R

rain forest A habitat where it rains almost every day. (page 50) Many kinds of plants and animals live in a rain forest.

recycle To make new items out of old items. (page 90) You can recycle paper.

reduce To cut back on how much you use something. (page 90) We should reduce the amount of water we use.

reflect To bounce off of something. (page 188) Light can reflect off shiny objects.

repel To push away or apart. (page 175) The two south poles of magnets repel each other.

reptile Animal with dry, scaly skin. (page 18) An alligator is a reptile.

reuse To use something again. (page 90) We can reuse things.

rock A natural resource that is hard and nonliving. (page 78) A rock is used as part of this tool.

rotation A turn or spin. (page 108) Earth makes one rotation in 24 hours.

S

seed Plant part that can grow into a new plant. (page 6) A seed can grow with water, sunlight, and air.

simple machine A tool that makes the force of a push or pull stronger. (page 168) This simple machine is called a ramp.

soil A mix of tiny rocks and bits of plants and animals. (page 82) Most plants need soil to grow.

solar system The Sun, eight planets, and their moons. (page 122) Planets in our solar system orbit the Sun.

solid Matter that has a shape of its own. (page 132) This chair is a solid.

solution A kind of mixture that is hard to take apart. (page 153) Water and drink mix make a solution.

sound A kind of energy that we can hear. (page 182) An alarm clock makes a loud sound.

speed How far something moves in a certain amount of time. (page 160) A cheetah has a fast running speed.

star An object in space made of hot gases that glow. (page 120) The Sun is a star.

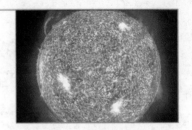

static electricity A kind of energy made by tiny bits of matter. (page 194) Static electricity makes the cat's fur stick to the balloon.

stratus Thin clouds that form into layers like sheets. (page 102) Stratus clouds can cover the sky.

 T

temperature How hot or cold something is. (page 94) A high temperature means something is warm.

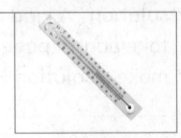

trait The way a plant or animal looks or acts. (page 13) Flower color is a trait.

 V

vibrate To move back and forth quickly. (page 182) Strings vibrate to make sound.

volcano An opening in Earth's crust and mantle. (page 74) A volcano can change the land quickly.

volume The amount of space something takes up. (page 137) You can measure the volume of a liquid with measuring cups.

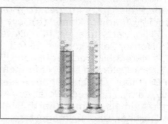

woodland forest A habitat where trees can grow well. (page 48) Many deer live in a woodland forest.

Credits